EATING MEAT

&

STAYING HEALTHY

EATING MEAT

&

STAYING HEALTHY

A post-nouvelle cookbook

JOSEPHINE BACON

BARRON'S

A QUARTO BOOK

First U.S. and Canadian edition published 1987 by
Barron's Educational Series, Inc.

Copyright © Quarto Publishing plc 1987

All enquiries should be addressed to:
Barron's Educational Series, Inc.
250 Wireless Boulevard
Hauppauge,
New York, NY 11788

International Standard Book No. 0-8120-5865-8

Library of Congress Catalog Card No:

Levy-Bacon, Josephine.
  Eating meat and staying healthy.

  (A Quarto book)
  Includes index.
  1. Cookery (Meat)   2. Meat.   I. Title.
TX749.L45 1987   641.6'6   87-14539
ISBN 0-8120-5865-8

This book was designed and produced by
Quarto Publishing plc
The Old Brewery
6 Blundell Street
London N7 9BH

Senior Editor Helen Owen
Art Editor Vincent Murphy

Designer Penny Dawes
Photographer David Burch
Home Economist Nicola Diggins

Art Director Moira Clinch
Editorial Director Carolyn King

Typeset by Burbeck Associates
Manufactured in Hong Kong by
Regent Publishing Services Ltd
Printed by Leefung-Asco Printers Ltd, Hong Kong

CONTENTS

MEAT IN YOUR DIET

Wendy Doyle

Following close on the heels of smoking, a diet rich in meat has come in for the brickbats of the health pundits. Containing protein, fat and cholesterol – all of which are judged to constitute an excessive proportion of today's intake – meat is no longer considered a daily necessity for a balanced diet. Indeed, the articles and books on vegetarianism that proliferate are positively discouraging on the healthy aspects of a meat-eating regime. But for the most part, these diatribes oversimplify the case. Throughout history humans have been *omnivorous*, not merely *carnivorous*, and the balance of a diet rich in meat, poultry, fish, vegetables and fruit has assured us of the coveted title of most successful species in the evolutionary stakes. Today, education and ever-wider options in food and drink should prevent us from succumbing to fads or scares and allow us to use our knowledge to guarantee a longer life for ourselves and a healthier, happier one for our children.

For the majority of us, to whom meat is still a normal, enjoyable part of eating, this means the adoption of a more responsible attitude. It means choosing animals – and cuts – that are as low as possible in saturated fat, and limiting our intake so that we do not "overdose' on the good things of life, including meat. The old-fashioned preference for meat and/or poultry at every meal is outmoded; neither should meat always be the focal point. Imaginative combinations with legumes, starches and vegetables will make less go further and provide contrasts of flavor and texture that the old "meat and potato" days could never match.

· A SOURCE OF GOOD THINGS ·

Once upon a time, we were hunters and gatherers. This meant we had the best of both worlds, for there is no denying the importance of vegetable foods in providing vitamin C, thiamine, carotene and many other important nutrients – as well as fiber – for our diet. On the other hand, it is equally true that trace elements, including iron, zinc, copper and selenium, were then, and still are, more readily absorbable when presented as animal products rather than in vegetable form. As for vitamin B12, it is simply not present in plants. Of course, people can live healthy lives as vegetarians, but it is much more difficult to obtain a balanced supply of nutrients from vegetable sources than from a mixture of plant and animal food. That is why mainstream nutritionists, doctors and sensible home cooks will always find a creative place for meat in their menus.

Most lean cuts of a particular meat or of poultry are equally nutritious, although the muscular sections are tougher because of the coarser muscle fibers and a higher proportion of connective tissue. This is why different methods of cooking are used: dry heat for tender roasts, steaks and chops, and moist heat for the tougher cuts.

· KEEPING A BALANCE ·

Today we are often advised to cut down on the amount of meat eaten in the interests of preventing heart disease. This is because fat – particularly saturated fat – is associated with domestically reared animals. It is interesting that in one of the predominantly meat-eating populations, the Eskimos, the incidence of heart disease is negligible, as are the types of cancer that are prevalent in North America and northern Europe. A similar picture emerges from meat-eating tribes like the Hadza of Tanzania, the Masai and El Molo of Kenya and many similar communities. These people obtain a high proportion of their energy from the flesh of wild species, which contains a large amount of lean meat compared with a small proportion of saturated fat. The combination of their outdoor, active life and diet low in fatty meat has resulted in a natural balance.

· REDUCING FAT INTAKE ·

It is worth noting that most people in developed countries consume an excess of energy. By reducing fat intake they can significantly reduce energy intake. This applies to fats from all sources, but animal fat is visible and avoidable. Fat consumption from meat can be reduced in a number of ways:

1. Select lean meat – go for quality rather than quantity
2. Trim visible fat
3. Cook without adding extra fat
4. Broil rather than fry
5. Skim off fat from stews and casseroles

· VARIETY – FOR GOOD HEALTH AND GOOD COOKING ·

No natural food is nutritionally complete, so it is important to eat a variety of different foods. Careful selection of accompanying dishes, as suggested with each recipe, will considerably reduce the percentage of energy you derive from fat – including saturated fat – as well as increasing the fiber content of your diet. The addition of pasta, rice and other cereals, preferably of the whole-grain variety, and of legumes, root vegetables, eggplants, peppers, etc. will boost not just your fiber intake but also your vitamin and mineral intake. It also makes economic sense in providing bulk, thereby diminishing the need for large, expensive portions of meat.

Not only do beef, lamb, pork and poultry look and taste different, but they also have certain different nutritional characteristics. For instance, the fat of ruminants – animals that chew the cud, including cattle, sheep and goats – tends to be more saturated than that of pigs and poultry.

BEEF Lean beef is a good source of iron and zinc. Consumer pressure for leaner beef is beginning to pay off, with the introduction of leaner breeds.

PORK AND BACON Pork traditionally contained more fat than other animals and this probably explains why it acquired the reputation of being indigestible. However, the greatest improvements in the production of leaner animals have occurred in the pig industry. Pork is considerably richer in thiamine (vitamin B1) than other meats and is also a good source of vitamin B6. Remember, pork should always be cooked thoroughly to avoid the risk of trichinosis.

LAMB Lean lamb is a relatively good source of vitamin B_6 and zinc. But it can be very fatty, so choose lean cuts, such as leg or noisettes from which the visible fat has been removed.

POULTRY Chicken and turkey without skin have the lowest fat content of any domesticated meat, white meat being slightly lower than dark meat. Removing the skin before cooking reduces the fat content by more than half. White meat also contains less connective tissue and is therefore particularly easy to digest. Duck and goose, in contrast, contain a considerable amount of fat in relation to protein and should be reserved for special occasions unless the skin and visible fat are removed *before* cooking. Always be sure to cook your poultry thoroughly, to avoid any chance of salmonella infection, a potential hazard of processing.

ORGAN MEATS The nutritional quality varies according to the type. Liver, kidney and heart are particularly good value for money, being rich sources of minerals and vitamins. They are also reasonably low in fat and the fat that *is* present is higher in the essential polyunsaturated fats than other parts of the animal. It is true that these are relatively high in cholesterol, but on a diet that is low in total fat, including saturated fat, you should be able to enjoy these foods without raising your blood cholesterol level. Liver, kidney and sweetbreads are especially rich in cells and so contain more nucleic acids than muscle. For this reason, patients with gout are usually advised to avoid them. Vitamins A, D and C, not present in lean tissue meat, are present in liver and kidney. These are also richer in vitamin B_{12}, folic acid and iron, all of which help to prevent anemia. At one time lightly cooked liver was the only means of saving the lives of patients with pernicious anemia. Tongue contains considerably more fat than other forms of organ meat.

GAME Game includes hunted wild birds and animals such as deer, hare and rabbit. The meat from these undomesticated animals is the kind eaten by our ancestors, when diseases of affluence – such as heart disease – were almost unknown. When these creatures are not reared by humans but have to rely on their own ability to find food, they are usually much leaner and fitter than domesticated animals and consequently provide a healthy alternative to our usual options of farmed meats.

· CONCLUSION ·

There has been so much talk about this food being unhealthy and that one being fattening that many of us have almost forgotten that eating should be a pleasure, not a compromise between good health and feelings of guilt. Healthy eating is an important part of staying well – though we must always remember that health and fitness depend not only on sensible eating habits but on a combination of factors, including exercise, nonsmoking, and the management of stress. It's common sense that prevention is better than cure, and in choosing this book you have taken a positive step in the direction of healthier living.

Wendy Doyle.

THE CUTS TO CHOOSE

Illustrated below are the leanest cuts of meat, as well as some popular favorites for
entertaining, such as crown roast of lamb. Always choose meat that looks moist
and fresh, and which has the least amount of visible fat. This will not only
give you more meat for your money, it will make far healthier eating for you
and your family.

The tables show the average grams of total fat per 100 grams of uncooked meat.
Trimming off any visible fat from red meat, and removing the skin from poultry
before cooking will considerably reduce the fat content, as the difference in the two
sets of figures given here shows.

The figures in the tables are taken from the *Handbook of the Nutritional Contents
of Foods*, US Department of Agriculture, 1975, and *Agriculture Handbook No. 8*,
US Department of Agriculture, 1979 and 1983. An asterisk indicates where figures
have not yet been published by the US Department of Agriculture.

BEEF

Boneless brisket flat half

Rump steak

Lean cubed chuck steak

Porterhouse steak

Round steak

Tenderloin

Rib-eye roast

Rib-eye steak

% *FAT*	Choice Grade		Good Grade	
	Trimmed	*Untrimmed*	*Trimmed*	*Untrimmed*
Chuck	8.0	25.3		
Club Steak	10.3	34.8	7.5	27.9
Ground	10.0 (Lean)	21.3 (Regular)		
Porterhouse	8.2	36.2	5.5	33.8
Round	4.7	12.3	*	*
Rib	11.6	37.4	*	*
Rump	7.5	25.3	5.4	21.4

Rolled brisket

When meat is prepared for **stewing**, the sinews and as much fat as possible should be removed.

Fillet and other cuts with a collar of fat should be trimmed of all visible fat.

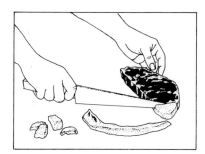

Steaks should be trimmed of fat before sautéeing or broiling.

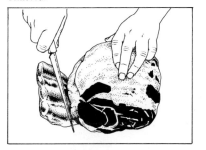

To prepare **roast rib**, remove the backbone before cooking. Using a long, sharp knife, cut and scrape as close to the bone as possible, leaving the thin rib bones.

To prepare a **beef roast** trim off excess fat, shaving it off neatly from the outside of the meat.

To **tie a roast**, roll the meat up like a jelly roll, and tie it neatly and firmly with string in several places.

VEAL

Escalopes

% *FAT*	"Thin" Class	"Medium-fat" Class
	Untrimmed	*Untrimmed*
Flank	18	27
Foreshank	5	8
Loin	8	11
Plate	12	17
Rib	9	14
Round with Rump	6	9

Rib chop

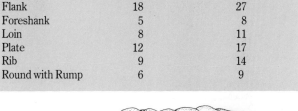

Boneless rump roast

Loin chop

Arm roast

Boned rolled shoulder

LAMB

Cutlets

Chump chops

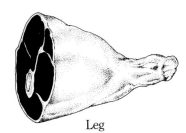

Leg

Boneless rolled leg

Rib chops

Crown roast

% FAT	Choice Grade		Good Grade	
	Trimmed	*Untrimmed*	*Trimmed*	*Untrimmed*
Leg	5.0	16.2	4.7	14.6
Loin	5.9	24.8	5.6	22.6
Rib	8.4	30.4	7.9	27.1
Shoulder	7.7	23.9	7.3	22.0

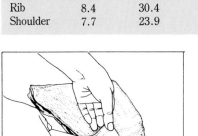

Two methods of preparing **rack of lamb** are shown here. For both, start by trimming off as much fat as possible from the skin side of the rack.

To serve **cutlets,** split rack of lamb into individual chops.

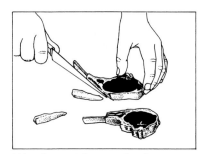

Trim off as much fat as possible. Cut off the exposed bone so that it is no longer than 1in.

To prepare a **crown roast**, you will need two racks. Trim away the strip of fat covering the bone ends, so they are exposed.

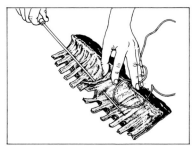

Lay the racks side by side, and tie them securely together in two places.

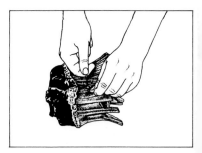

Curve the two joined racks round into a ring, and tie the ends securely together in two places.

PORK AND HAM

Whole loin

Spareribs

Blade Boston roast

Fillet

% FAT	Trimmed	Untrimmed
Leg (Ham)	5.4	20.8
Loin	7.5	24.1
Tenderloin	2.5	–
Spare ribs	–	23.6

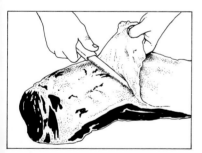

Loin chops

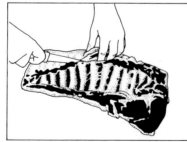

Butt end gammon

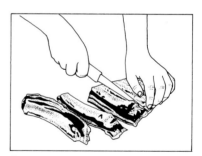

Ham hock

To prepare **spareribs,** pare away the skin and visible fat with a sharp knife.

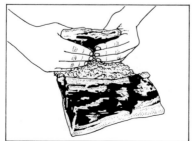

Slice off the fat part extending beyond the rib tips.

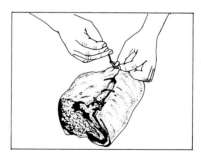

Cut out each bone and trim neatly, removing as much fat as possible.

To prepare **rolled stuffed loin**, trim away excess fat on the underside of the meat.

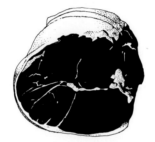

Raise the flap to cover the stuffing, and roll up the meat like a jelly roll.

Tie securely with string in several places, so the stuffing is held firmly in place.

13

POULTRY

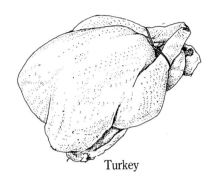

Turkey

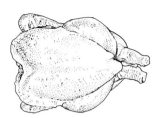

Roasting chicken

Skinned breasts of chicken

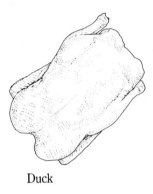

Duck

Rock Cornish hen

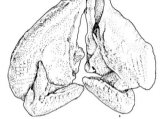

Wing and breast of chicken

Chicken legs

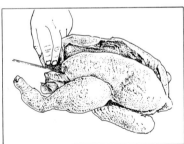
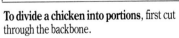

To divide a chicken into portions, first cut through the backbone.

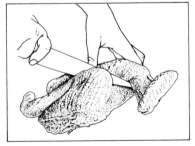

Insert knife blade into the skin between the body and thigh. Pull the thigh away.

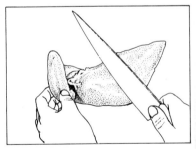

Skin the portions, and scrape away visible fat and membranes.

% FAT	Chicken	Without Skin	With Skin	Turkey	Without Skin	With Skin
	Roasting	2.7	15.8	Roaster/Fryer	1.6	4.2
	Light Meat	1.6		Light Meat	0.5	3.8
	Dark Meat	3.6		Dark Meat	2.7	4.8
	Broiler/Fryer	3.1	15.1			
	Light Meat	1.5		Duck	6.0	39.3
	Dark Meat	3.8		Goose	7.1	33.6

GAME

Rabbit

Partridge

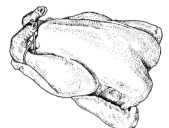

Pheasant

% *FAT*	*Without Skin*	*With Skin*
Guinea Fowl	2.5	6.5
Pheasant	3.6	9.3
Rabbit (Tame)	8.0	
(Wild)	5.0	
Venison	4.0	
Pigeon (Squab)	7.5	23.8

Quail

Venison

VARIETY MEATS

Pig's liver

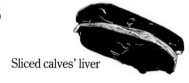

Sliced calves' liver

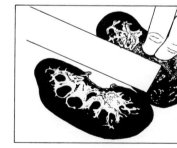

Trim **kidneys** by removing the white core, and as much fat as possible.

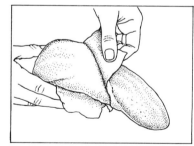

The skin will peel away from a **beef or ox tongue** easily after cooking.

Ox tongue

Chicken livers

Kidneys

		% *FAT*
Liver:	Calves'	4.7
	Chicken	3.7
	Lamb's	3.9
	Pig's	3.7
Tongue:	Calves'	5.3
	Lamb	15.3
	Ox	15.0
Gizzard:	Chicken	2.7

CHOOSE THE RECIPE TO SUIT YOUR DIET

**While all the recipes in this book have been compiled
with healthy eating in mind, the following dishes are especially suitable for
those with particular dietary preferences.**

LOW CALORIE RECIPES
Less than 225 calories per serving

	page		page
Lemony Meatball Soup	77	Chicken Tikka (excluding Raitas)	118
Wedding Soup	96	Chinese Chicken Salad with Sesame Seeds	136
Caribbean Spareribs with Fruity Sauce	60	Chicken Packets with Spring Vegetables	120
Drip-Roast Veal with Coriander	61	Spiced Rock Cornish Hens with Fresh Fruit Sauce	120
Veal Scallops with Sage and Lemon	100	Caribbean Chicken	115
Light Ham Mousse	92	Marinated Chicken Brochettes	116
Stir-Fried Lean Pork with Chinese Vegetables	117	Deviled Turkey Legs with Puréed Vegetables	141
Lemony Chicken-in-the-Pot	109	Light Chicken Liver Pâté	170
Clay Pot Chicken with Olives			

LOW FAT RECIPES

No more than 1.5g saturated fat per serving		*No more than 6g total fat per serving*	
Lemony Meatball Soup	32	Lemony Meatball Soup	32
Beef Napoleon	37	Wedding Soup	77
Veal Piccata with Tagliatelle Verdi	60	Barbecued Ribs with Orange-Mustard Sauce	45
Lamb Pizza with Pine Nuts	76	Veal Piccata with Tagliatelle Verdi	60
Creole Beans and Rice with Smoked Ham	101	Drip-Roast Veal with Coriander	60
Light Ham Mousse	100	Creole Beans and Rice with Smoked Ham	101
Caribbean Spare Ribs with Fruity Sauce	96	Light Ham Mousse	100
Chicken Tikka (excluding Raitas)	118	Caribbean Spareribs with Fruity Sauce	96
Chicken Pasta Primavera	128	Chicken Salad with Special Low-Fat Mayonnaise	114
Chicken Packets with Spring Vegetables	120	Marinated Chicken Brochettes	116
Turkey Tonnato	149	Chicken Tikka (excluding Raitas)	118
Deviled Turkey Legs with Puréed Vegetables	141	Chicken Packets with Spring Vegetables	120
Turkey Montmorency	145	Braised Chicken with Brown Rice and Raisins	108
Braised Turkey with Sweet Potatoes and		Turkey Tonnato	149
Brussels Sprouts	148	Deviled Turkey Legs with Puréed Vegetables	141
Haunch of Venison with Wild Mushrooms	160	Braised Turkey with Sweet Potatoes and	
Roast Pigeon with Buckwheat Groats	164	Brussels Sprouts	148
Roast Pheasant with Blackberry and Apple	161		

LOW CHOLESTEROL RECIPES
Less than 60mg Cholesterol per serving

Lemony Meatball Soup	32	Festival Pilaf with Ground Lamb	72
Wedding Soup	77	Lamb Pizza with Pine Nuts	76
Persian Meatballs with Spinach	48	Braised Pork and Black Beans with Oranges	89
Stir-Fried Beef with Peppers and Mangetout	53	Creole Beans and Rice with Smoked Ham	101
Raw Lamb Pâté with Cracked Wheat	81	Light Ham Mousse	100

Caribbean Spareribs with Fruity Sauce	96	Braised Guinea Fowl with Mushrooms and Bilberries	157
Stir-Fried Lean Pork with Chinese Vegetables	92	Roast Pheasant with Blackberry and Apple	161
Chinese Chicken and Walnuts	128		

LOW SALT RECIPES

Less than 150mg sodium per serving

Wedding Soup	77	Barbecued Chicken with Grapes	125
Stir-Fried Beef with Peppers and Mangetout	53	Marinated Chicken Brochettes	116
Barbecued Ribs with Orange Mustard Sauce	45	Chicken Tikka (excluding Raitas)	118
Beef Java	44	Chicken Packets with Spring Vegetables	120
Cold Spiced Beef	33	Jaffa Roast Chicken	105
Paprika Beef	32	Fasenjan Chicken	132
Beef Teriyaki	52	Baked Chicken with Prunes	133
Kleftiko	84	Spanish Chicken and Brown Rice	113
Mexican Pork Steaks	92	Caribbean Chicken	115
Stir-Fried Lean Pork with Chinese Vegetables	92	Chicken Breasts with Green Peppercorns	119
Braised Pork and Black Beans with Oranges	89	Chinese-style Spiced Braised Duck	152
Caribbean Spareribs with Fruity Sauce	96	Broiled Liver with Fresh Herbs	169

HIGH FIBER RECIPES

More than 8g fiber per serving

Cabbage Leaves with Beef and Almond Stuffing	36	Lamb Pizza with Pine Nuts	76
Medallions of Veal with Spring Vegetables	64	Braised Pork and Black Beans with Oranges	89
Festival Pilaf with Ground Lamb	72	Creole Beans and Rice with Smoked Ham	101
Lamb Couscous with Seven Vegetables	69	Chicken Salad with Special Low-Fat Mayonnaise	114
Minted Lamb Pie	80	Saddle of Rabbit with Prunes and Apricots	165

NUTRITIONAL INFORMATION

Calories, fat, saturated fat, cholesterol, sodium and fiber contents are given per single serving of each recipe, excluding accompaniments. Each recipe has also been given a merit rating, indicating the benefits of low-fat, high-fiber dishes: the top recipes win three stars. The star ratings show how *low* the fat, saturated fat, cholesterol and sodium contents are (★★★), and how *high* the fiber content is (★★★). The scale is relative, and the recipes in this cookbook all tend to be lower in fat, saturated fat, cholesterol and sodium than equivalent recipes in other books. The bandings used are set out below:

		LOW Less than:	MEDIUM	HIGH More than:
Calories	kcals	250	250 – 500	500
Fat	g	8	8 – 16	16
Saturated fat	g	2	2 – 5	5
Cholesterol	mg	75	75 – 200	200
Sodium	mg	200	200 – 400	400
Fiber	g	3	3 – 8	8

If you include the accompaniments suggested for each recipe, you will also obtain a good fiber intake.

INTRODUCTION TO THE RECIPES

Josephine Bacon

A universal message from modern dieticians is that meat should be eaten in moderation. The days are gone forever when it was considered a sign of affluence to eat meat three times a day – once a day is more than enough. Eating meat in "extended" form, that is, making it go further by mixing it with starches and legumes, is becoming increasingly popular. In countries where this is the common practice, it has been noticed that the kind of dietary diseases we suffer from in the West are almost nonexistent. Adaptations of the foods of these countries have been included in the book, and their cooking methods have been adopted. The recipes themselves are designed to show you how to get the best out of the meat you eat, and the dishes are as low as possible in fats, particularly in saturated fat and sodium.

· FATS AND OILS ·

Even a cursory glance through classic cookbooks – including those written as recently as the 1970s – reveals that, by today's standards, meat dishes were cooked using an excessive amount of fats and salt. Roasts were basted in their own cooking fat; game and veal were larded (threaded with strips of pork fat) and barded (trussed with a sheet of pork fat), and even casseroles and stews were flavored with slabs of greasy, salty pork. Suet was used lavishly in both sweet and savory dishes.

The excuse for excessively fatty cooking was that people needed the extra energy, as they did heavy, manual labour. Then, too, in colder climates home heating was very inadequate. Neither condition applies today. In warmer climates, where other types of fat are used, traditional cooking still tends to be greasy.

All kinds of medicinal properties have been attributed to olive oil. Whatever the merits and demerits of these claims it should, like all oils and fats, be used sparingly. The recipes in this book do not specify a particular type of oil unless the success of the dish depends upon its inclusion. Generally speaking, oils high in polyunsaturates, such as grape seed oil and safflower oil, are by far the best. These can now be bought at supermarkets and health food shops, and they are finding their way into many local food shops. The other popular cooking oils include wheat germ, soy, corn, sunflower, peanut and, finally, olive oil. Low-fat spreads (try to avoid those made with cottonseed oil) and butter substitutes made with buttermilk are not suitable for frying, since they have a high water content and are likely to splatter. However, they can be used for all other cooking methods.

· UTENSILS AND KITCHEN EQUIPMENT ·

Of course, many people simply like the taste and texture of oils and fats, but there is another reason such large quantities have traditionally been used in cooking. Oil and grease are lubricants, and just as mineral oil and axle grease make machine parts run smoothly, so vegetable oil and butter lubricate pots and pans and keep food from sticking to them. Much of the fat and grease used in traditional recipes was there simply to help the cook, not the diners. Today's cooking utensils are designed to prevent food from sticking to the pan, and the modern cook should make full use of them to promote healthier eating.

In addition to nonstick pans, it is sometimes forgotten that microwave and combination ovens fall into this category. These ovens cook food from the inside out, by agitating the molecules inside the food. This means the food "sticks to itself," rather than to the dish in which it is cooked. It is quite unnecessary to grease dishes used to cook food in a microwave oven, and meat dishes will be perfectly moist and delicious even when completely devoid of added fat.

Not all meat dishes are suitable for microwave cooking; those that require slow cooking, in order to break down the fibers, should be cooked conventionally. Stews can also be cooked in the special slow-cooking pots, providing the ingredients have been brought to the boil for 5 minutes on top of the stove, before being transferred to the pot. Cooking food from scratch in a slow-cooking pot can allow harmful substances that are usually killed during the conventional cooking process to persist in the food.

Another aid to better meat cooking is the food processor. Although experienced cooks chop meat with two knives, before the advent of the food processor most of us relied on the grinder. This now-obsolete piece of equipment squeezed out most of the meat juices, making hamburger, for example, dry and tough. The food processor method of grinding retains all the food's original juices.

· THE USE OF SALT ·

In the past, too much salt was often added to meat dishes, and the Western palate has become used to highly salted food as a result. The reason for this was that salting meat and fish was the best way to preserve it. Modern food preservation methods do not involve salt, although salted meats, such as ham and bacon, have remained popular. More than enough sodium is naturally present in meat and other foods for nutritional purposes, so we do not need to add any at all to our food, especially in temperate climates. In fact, salting meat before cooking breaks down the cell walls, allowing the blood to flow out and the fibers to toughen, so the meat becomes less tasty and tender. The amounts of salt added to food during cooking have been kept to a minimum in these recipes. Other flavors, such as herbs, hot pepper and sour flavors, have been substituted for salt, wherever possible. For those who prefer saltier food, a little salt can be added after cooking. Salt substitutes are discussed in the Standard Preparations section (see p.26).

· TRIMMING AND PREPARATION ·

Meat should always be trimmed of all visible fat, and large cuts often need further preparation, such as boning. This preparation has been kept to a minimum, and where it is done, it is always on the kind of meat you would serve at a dinner party or on a special occasion. Cooks nowadays do not want to waste time on elaborate preparations for everyday meals.

Great efforts have been made to make the recipes in this book as varied as possible. Every type of meat has been used, and every type of cooking method is represented – except, of course, deep frying. Low-fat milk, yogurt and spreads have been used whenever possible in these recipes and, to help the healthy eater avoid the lure of the deep-fat fryer, accompaniments are suggested with each dish to make up a balanced meal. I hope you find this range of meat dishes not only healthier but far tastier than the fatty, greasy meat dishes of the past.

Josephine Bacon

THE RECIPES

A full explanation of the nutritional information in this book appears on page 17. The star ratings for the recipes are based on the following figures.

		LOW *Less than:*	MEDIUM	HIGH *More than:*
Calories	kcals	250	250 – 500	500
Fat	g	8	8 – 16	16
Saturated fat	g	2	2 – 5	5
Cholesterol	mg	75	75 – 200	200
Sodium	mg	200	200 – 400	400
Fiber	g	3	3 – 8	8

RECIPE NOTES

All spoon measures are level and all cup measures are packed.
Flour is all-purpose unless otherwise stated.

STANDARD PREPARATIONS

The following standard preparations should be used where specified in the recipes but also make useful recipes on their own. Broth made from scratch will be less fatty than bouillon cubes, which need the fat to bind them together, and will not contain the artificial flavorings and colorings that are contained in so many processed foods. All the broths can be frozen if well strained, and they will keep for months in the freezer; if refrigerated, use within 1 week of making. One of the broths, served with one of the breads, will make a great meal in itself, especially if you add a salad.

BEEF BROTH

Marrow bones contain nutritious but fatty bone marrow. Chilling and skimming the fat from the surface of the broth will allow the nutrients in the marrow to be absorbed during cooking but enable the fat to be discarded before eating. This recipe gives you a good piece of boiled beef in addition to the nutritious broth.

· METHOD ·

Put the bones into a large stockpot and add the meat and 3½ quarts water. Bring to the boil, skimming off the scum that rises to the surface. When the water boils, add 6 tablespoons cold water to the pot and skim again; repeat the operation once more.

Skim until there is hardly any scum rising to the surface and the broth is clear.

Add the vegetables and skim off any more scum. With the lid of the pot slightly askew, continue simmering the contents gently for 3 hours.

Remove the meat and bones, discarding the latter. Strain the liquid through a colander or sieve lined with two layers of cheesecloth.

Allow the liquid to cool, then use a metal spatula to skim off the fat that congeals on the surface. Chill the broth until required. It will keep for 5 days in the refrigerator and up to 3 months in the freezer.

· SERVING SUGGESTION ·

This broth is excellent as a beef soup served on its own or with the addition of vegetables, such as carrots and peas. The meat can be thinly sliced and eaten cold with salad, or reheated and served with as large a variety of small vegetables as are available (potatoes, onions, carrots, peas, beans, etc.).

INGREDIENTS

2lb beef marrow bones

2lb round rump roast, trimmed of visible fat, rolled and tied with string

3 carrots, split lengthwise

2 turnips, split in half

2 leeks, split lengthwise

1 bay leaf

1 teaspoon dried thyme

1 parsnip, split lengthwise

½ teaspoon salt or salt substitute (page 26)

Makes 2½ quarts

Calories	14	★ ★ ★
Fat	0.5g	★ ★ ★
Saturated Fat	0.2g	★ ★ ★
Cholesterol	6mg	★ ★ ★
Sodium	86mg	★ ★ ★
Fiber	0.1g	★

Based on 10% extraction in cooking

INGREDIENTS

*6½lb lamb bones (shank
      and bone-in neck slices)*

2lb veal knuckle or beef shin

1 leek, sliced lengthwise

3 onions, 1 stuck with 2 cloves

1 parsnip, split lengthwise

2 turnips, halved

4 large carrots, split lengthwise

*½ teaspoon salt or salt substitute
      (page 26)*

Makes 3 quarts

Calories	20	★ ★ ★
Fat	1g	★ ★ ★
Saturated Fat	0.3g	★ ★ ★
Cholesterol	9mg	★ ★ ★
Sodium	16mg	★ ★ ★
Fiber	0.1g	★

Based on 10% extraction in cooking

LAMB BROTH

Lamb broth is always fattier than chicken or beef broth, and this affects its keeping properties, so use it fairly soon after making. It makes an excellent base for both veal and lamb dishes.

· METHOD ·

Trim all visible fat from the meat. Put the bones into the pot first, then add the meat. Add 5 quarts water, and bring to the boil slowly, skimming off any scum that rises to the surface with a slotted spoon. When the liquid starts to boil, add 6 tablespoons cold water and continue skimming until the liquid boils again. Repeat this operation twice more, or until there is hardly any scum and the broth is clear. Add the remaining ingredients and bring back to the boil. Partially cover the pot and simmer for 5 hours, adding 6 tablespoons cold water once every hour.

Remove and discard the bones, and reserve the meat. Discard the rest of the ingredients and strain the broth through a strainer or colander lined with two layers of cheesecloth. Allow the broth to cool, then skim off the fat that congeals on the surface with a metal spatula. Chill until required. The broth will keep for 5 days in the refrigerator and up to 3 months in the freezer.

· SERVING SUGGESTION ·

The lamb can be served Victorian-style with capers or caper sauce (made with skimmed milk and thickened only with cornstarch). You will be surprised at its tenderness and flavor; boiled lamb's unsavory reputation is quite undeserved. The broth and meat can be used as the basis for a stew, by adding cubes of the cooked lamb and vegetables to the liquid. If you are making a small quantity of stew, it can be cooked quickly and deliciously in a microwave oven.

CHICKEN BROTH

Always try to use giblets in chicken broth to improve the flavor. A whole chicken that has been used to make broth may not have enough flavor left for eating whole, but it would still be fine for a cooked, ground chicken recipe. Chicken feet are gelatinous and are also very good for broth. The removal of the skin and surplus fat before cooking will not affect the flavor, because the true flavor of all meat broths comes from the vegetables and seasonings added to the meat.

INGREDIENTS

*2lb chicken or the equivalent
      weight in carcasses, plus giblets*

1 turnip, quartered

1 onion, stuck with 2 cloves

2 celery stalks

1 large carrot, split lengthwise

1 cup parsley

*½ teaspoon salt or salt substitute
      (page 26)*

Makes 2 quarts

· METHOD ·

If using a whole chicken, skin it and remove all visible fat. Put the chicken or carcasses into a large pot and add 4½ quarts water. Bring to the boil, uncovered, skimming off any scum that rises to the surface. Reduce the heat, cover the pot and simmer for 1½ hours. Add the rest of the ingredients and simmer for a further 1½ hours.

Strain off the liquid and leave it to cool. When cold, use a spoon or metal spatula to skim any further fat from the surface. The broth will keep for 5 days in the refrigerator and for up to 3 months in the freezer.

· SERVING SUGGESTION ·

This broth can be served as chicken soup or clear consommé. Sprinkle it with chopped parsley or other fresh herbs (lemongrass or rosemary will give it an unusual flavor) before serving.

(per 250ml)		
Calories	16	★ ★ ★
Fat	1g	★ ★ ★
Saturated Fat	0.1g	★ ★ ★
Cholesterol	11mg	★ ★ ★
Sodium	112mg	★ ★ ★
Fiber	0.1g	★

Based on 10% extraction in cooking

MIXED MEAT BROTH

This all-purpose broth can be made with any leftover meat, cooked or uncooked. The meats listed below are just a guideline; you can make it with any combination of meats, provided you stick roughly to the total weights given here. For instance, you can sometimes get chicken necks at a bargain price from the butcher, and you could use all necks for the chicken ingredient.

· METHOD ·

Remove all visible fat from the meat. Put the meat into a large stockpot, add 5 quarts water and bring slowly to the boil, skimming with a slotted spoon to remove any scum. When the water reaches the boil, add the remaining ingredients, cover the pot and reduce the heat. Simmer on very low heat, with the pot partially covered, for 3 hours.

Remove the beef and veal and reserve them for eating. Discard the bones and chicken parts. Line a strainer or colander with two layers of cheesecloth and strain the broth into a large bowl. Cool to room temperature, then refrigerate until cold. Skim off the congealed fat with a metal spatula.

The broth will keep for 5 days in the refrigerator and for up to 3 months in the freezer.

· SERVING SUGGESTION ·

To serve the beef, let it cool, then slice it thinly and serve with spring vegetables (baby carrots, brussels sprouts, snap beans, etc.).

The broth can be served as consommé and makes a nourishing light meal with whole wheat bread or toast.

INGREDIENTS

2lb beef marrow bones
2lb lean beef cuts (chuck, blade, flank steak)
2lb chicken pieces (backs, necks and feet)
1 leek, split lengthwise
1 bay leaf
1 onion, stuck with 2 cloves
1 garlic head
4 large carrots
1 parsnip, split lengthwise
4oz parsley
½ teaspoon salt

Makes 3 quarts

(per 250ml)		
Calories	22	★ ★ ★
Fat	1g	★ ★ ★
Saturated Fat	0.3g	★ ★ ★
Cholesterol	12mg	★ ★ ★
Sodium	47mg	★ ★ ★
Fiber	0.2g	★

Based on 10% extraction in cooking

INGREDIENTS

1 large egg
½ teaspoon garlic powder
½ teaspoon salt
½ teaspoon dry mustard
2 cups plain yogurt

Makes about 2 cups

Calories	9	★ ★ ★
Fat	0.25g	★ ★ ★
Saturated Fat	0.1g	★ ★ ★
Cholesterol	5mg	★ ★ ★
Sodium	23mg	★ ★ ★
Fiber	0g	★

per 15ml tablespoon

INGREDIENTS

1 grapefruit (rind only)
1 tablespoon ground allspice
½ tablespoon citric acid crystals

Makes 3 tablespoons

SPECIAL LOW-FAT MAYONNAISE

**I discovered long ago that an absolutely delicious mayonnaise
can be made by replacing all that oil in mayonnaise
with plain yogurt. You can try different kinds, such as sheep's or
goat's milk yogurt, and if you want the mayonnaise to be especially
low in fat, use low-fat yogurt.
You can also vary the spicing by substituting chili
powder for the mustard, for instance. This mayonnaise keeps well
when chilled (do not freeze it), and is extremely simple to make if
you have a food processor or blender.**

· METHOD ·

Put the egg and flavorings into a blender or food processor, or beat with a whisk until fully combined. Gradually add the yogurt with the machine running, or while beating, if you are using a whisk. Continue beating or whisking until the mixture is thick and homogeneous.

· VARIATIONS ·

Substitute ground ginger for the mustard and use onion powder for the garlic powder – this variation has a subtle taste that combines well with Indian curries.

Add a tablespoon of any chopped fresh herbs that are in season. Mint or rosemary is perfect for lamb dishes; sage goes well with pork, and coriander complements veal or beef recipes.

Use a pinch of ground turmeric to give the mayonnaise an attractive golden color.

Grate the rind of one lemon into the mixture, or add a tiny pinch of citric acid crystals, for a fresh, zesty taste.

SALT SUBSTITUTE

**This homemade salt substitute relies on the principle that a very
sour flavor is a good substitute for a salty one. You can use dried
lemon or orange rind instead of the grapefruit,
but the flavor will not be as strong.
Citric acid crystals – also called lemon salt or sour salt – can be
bought from delicatessens specializing in Jewish
and eastern European foods.**

· METHOD ·

Peel the grapefruit as thinly as possible. Scrape away all the white parts. Dry the rind for 8 hours or overnight over a hot radiator, in a shaded place outdoors, or in a gas oven with just the pilot light lit.

Grind the rind in a coffee grinder or spice grinder. Combine it with the other ingredients. Put it into a well-sealed bottle and shake well to mix. Store in a dry place.

· VARIATIONS ·

A tablespoon of freshly ground black pepper added to the mixture will make it into citrus pepper, ideal for seasoning meat. If ground ginger is substituted for the pepper and lemon for the grapefruit, it makes a very good beef seasoning.

· OTHER SALT SUBSTITUTES ·

The baking powder, used mainly in German and Swedish baking, which is the ancestor of our own baking powder consists of potassium carbonate. It is also known as hartshorn, because it was made from burned, powdered horn. It still makes an excellent salt substitute. There are several salt substitutes on the market, but some contain chemical additives and colorings that are potentially allergenic, so read the labels carefully before buying.

Nutrient content negligible per portion

CREOLE SPICE

This seasoning complements the flavor of all kinds of meat, as well as adding an attractive coloring, especially to meat that is not roasted (such as microwaved or stewed meat).
Almost all spice mixtures of this kind contain large quantities of salt; "lemon pepper," for instance, often lists salt first in the ingredients, which means that is the *main* ingredient! Making your own mixture is the best way to ensure you are not eating hidden salt.

INGREDIENTS

2 tablespoons citric acid crystals
2 tablespoons garlic powder
1 tablespoon ground bay leaves
1 tablespoon cayenne pepper
1 tablespoon chili powder
2 teaspoons ground dried thyme
2 teaspoons ground dried basil
2 teaspoons ground black pepper
1 teaspoon ground coriander

Makes 4oz

· METHOD ·

Put all the ingredients into a small jar with a tight seal. Shake to mix well. Store in a dry place away from sunlight.

VARIATION For exotic East Indian dishes, substitute garam masala and curry powder for the thyme and basil.

Nutrient content negligible per portion

INGREDIENTS

*13 cups stone ground whole wheat
bread flour*

1 teaspoon sea salt

*2 tablespoons fresh yeast or 1 package
dry yeast*

5 cups tepid water

1 tablespoon honey

Oil for greasing the pans

Makes 3 loaves (each serving 8)

Picture: page 99

WHOLE WHEAT BREAD DOUGH

**Some salt is essential in yeast baking, since it inhibits the yeast
from rising too much. In addition, since whole wheat bread
contains bran and wheat germ oil it does not require much
kneading, so it is quick and easy to make.
Remember that all utensils should be warm when making yeast
mixtures. If you are using dry yeast, read the
manufacturer's instructions first.
If they recommend combining the yeast directly with the flour, do
so, and add the honey and liquid in the order they suggest.**

· METHOD ·

Mix the flour and salt. Combine the yeast with 1 cup of the water and the
honey. Cover with a damp cloth and leave in a warm place until the yeast
foams, about 20 minutes. Add the yeast mixture to the flour and pour in
the rest of the water. Stir well with a wooden spoon until the mixture
leaves the sides of the bowl cleanly.

Brush three 1-quart nonstick loaf pans lightly with oil and put them in
a warm place. Divide the dough, and put into the pans. Cover with a damp
cloth and leave in a warm place to rise (45 minutes).

Bake the loaves in a preheated 400°F oven for 40 minutes. Cool
slightly before removing from the pans. When cold, the bread can be
frozen until required.

· VARIATIONS ·

SEED-TOP LOAVES: Just before baking, brush the top of the loaves with
skimmed milk, and sprinkle them with sesame and/or poppy seeds just
before baking. Before rising, mix dried herbs into the mixture if desired.
The crunchy finish to the bread makes a particularly delicious complement
to pork and ham dishes.

COTTAGE LOAVES: Divide the dough into four pieces, two of them twice
the size of the other two. Sprinkle a baking sheet with semolina or
cornmeal. Place the 2 larger pieces on the baking sheet, and place a
smaller piece over each. Bake the two cottage loaves for one hour, or until
they sound hollow when rapped with the knuckles. This is a protein and
fiber-rich version of a traditional favorite.

FLAT BREADS: Roll out the dough into flat rounds using a rolling pin.
Sprinkle the surface generously with whole wheat flour. Arrange the
rounds between sheets of waxed paper and refrigerate overnight. The
next day, quickly brown the breads on a hot griddle or under a broiler.
Serve hot.

WHOLE WHEAT PIZZA: Divide the dough into 10 pieces before rising.
Flatten each piece into a circle with a well-floured rolling pin. Lay the
circles on oiled cookie sheets, cover and let rise for 40 minutes. Arrange
the pizza ingredients on one or more of the circles and bake at the oven's
hottest setting for 25 minutes. The dough circles can be frozen as is, or
baked as pizzas and then frozen until reheated. They are a kids' favorite!

Calories	203	★★★
Fat	1g	★★★
Saturated Fat	0.2g	★★★
Cholesterol	0mg	★★★
Sodium	85mg	★★★
Fiber	6.1g	★★

WHITE BREAD DOUGH

**Contrary to popular belief, white flour is not unhealthy –
particularly if you use unbleached enriched flour, which has a lot of
added vitamins, although it lacks the bran and wheat germ oil of
whole wheat flour. However, white flour doughs taste better
with some dishes.
If you are using dry yeast, read the manufacturer's
instructions carefully before starting.
If they recommend mixing the yeast directly with the flour before
adding liquid, do so and add the water later.**

· METHOD ·

Dissolve the yeast in 6 tablespoons lukewarm water; if you are using dry yeast, add the honey. Mix in 2 cups of the flour, kneading the mixture lightly. Cover and leave it to rise until doubled in bulk, about 30 minutes. Beat in the rest of the flour, $2\frac{1}{2}$ cups lukewarm water and the salt. Knead the dough for 5 to 10 minutes, until it is smooth and elastic and no longer sticks to your hands or the bowl.

Divide the dough in two and shape each half into an oblong loaf. Sprinkle a baking sheet with a generous layer of semolina or cornmeal. Lay the loaves on the sheet and cover them lightly with a dampened cloth. Leave them in a warm place to rise for about 2 hours. Just before baking, use a very sharp knife or razor blade to make a slash along the length of each loaf.

Bake the loaves in a preheated 400°F oven for 40 minutes, or until the loaves are golden brown and sound hollow when rapped on the bottom.

· VARIATIONS ·

The loaves can be patted into many different shapes, and for extra nutrition – if they are for children's food, for instance – replace the water with low-fat milk or add skimmed milk powder to the water.

EAST INDIAN BREAD (NAAN): Divide the dough into 10 pieces before rising. Flatten each piece into a circle with a well-floured rolling pin. Transfer to oiled baking sheets and let rise, covered, in a warm place for 40 minutes. Brush with water or low-fat milk and sprinkle each circle with a teaspoon of mixed sesame seeds, black onion seeds and nigella seeds. Bake at the hottest setting for 15 minutes.

PIZZA: Follow the instructions for Whole Wheat Pizza.

INGREDIENTS

*2 tablespoons fresh yeast or 1 package
dry yeast*
1 teaspoon honey (optional)
8 cups unbleached enriched white flour
1 teaspoon sea salt
Semolina or cornmeal
Extra flour for sprinkling

Makes 2 loaves (each serving 10)
Picture: page 99

Calories	181	★ ★ ★
Fat	1g	★ ★ ★
Saturated Fat	0.1g	★ ★ ★
Cholesterol	0mg	★ ★ ★
Sodium	100mg	★ ★ ★
Fiber	1.9g	★

BEEF

Always look for meat that is bright red in color and juicy looking. Marbled beef is fattier, but more tender; for this reason, the larger the amount of marbling the higher the USDA grading. There are four retail USDA grades: prime, choice, good and standard; prime and choice are the most flavorful cuts, but good and standard are fine for slow cooking, and, of course, they are leaner.

Beef is always hung to tenderize the fibers, and this is done in cold storage or in transit after slaughtering, depending on how far the meat has had to travel. Beef must be refrigerated as soon as possible after purchase. It is best stored in the coldest part of the refrigerator; do not freeze beef unless it is a large piece and in very good condition. Try to eat the beef within 1 or 2 days of purchase.

You will need between 4 and 6oz per serving for boneless beef, and about 1lb per serving for beef on the bone.

LEAN TERRINE

Most classic terrine recipes use lots of barding fat, butter or fat bacon to make them juicy. This one, although based on a traditional British recipe, needs no added fat provided it is cooked in a nonstick loaf pan or in a microwave oven. Microwaving is really a kind of dry steaming in which the water contained naturally in all food is turned into vapor and so cooks the food; for this reason it is the ideal way to cook dishes – such as custards and pâtés – that require water-baths in conventional cooking. All the new combination convection and steam ovens are also most suitable for this type of dish.

· METHOD ·

Blanch the cabbage leaves in boiling water for 5 minutes to wilt them.

Combine all the remaining ingredients, except the paprika, in a bowl. Feed them little by little into a food processor fitted with the metal blade, with the machine running, and continue to grind them for a couple of minutes longer.

Use the cabbage leaves to line a large, nonstick loaf pan if the terrine is being baked conventionally, or a glass container, if it is being cooked in a microwave or convection oven. Press the mixture down firmly over the cabbage leaves and smooth the top. If using a conventional oven, cover the terrine with several layers of nonstick baking paper and tie them down securely. Put the pan into a water bath so that the water comes about two-thirds of the way up the sides of the pan. Bake in a preheated 375°F oven for 3 hours, or until the mixture is firm to the touch and a skewer inserted into the center comes out hot. Top up the bath with boiling water every half-hour.

If the terrine is being cooked in a microwave, convection or combination oven, cover the glass container with plastic wrap. Do not vent the plastic wrap. Cook for 25 minutes on medium (or follow manufacturer's recommendations for pâtés).

Cool the terrine in the dish, then refrigerate overnight. Unmold before serving. Slice and sprinkle the slices with paprika.

· SERVING SUGGESTION ·

This is an ideal light luncheon dish or a dinner party starter because it can be made well in advance. Serve with a salad. For special occasions, the slices can be arranged on a serving dish and decorated with raw, sliced vegetables.

INGREDIENTS

6 large green cabbage leaves

1lb lean ground beef

8oz lean ground ham

2¼ cups fresh whole wheat
    breadcrumbs

½ teaspoon salt

1 teaspoon grated lemon rind

1 teaspoon cayenne pepper

¼ teaspoon grated nutmeg

½ teaspoon paprika

6 servings

Picture: page 34

Calories	272	★ ★
Fat	7g	★ ★ ★
Saturated Fat	2.6g	★ ★
Cholesterol	81mg	★ ★
Sodium	877mg	★
Fiber	4.7g	★ ★

INGREDIENTS

2lb leeks, trimmed, white parts
   only

2 tablespoons diet margarine

Juice of 1 lemon

2 cups chopped celery

2 quarts beef broth (page 25)

½ cup coarsely chopped parsley

FOR THE MEATBALLS

12oz cooked or raw lean ground
   beef

1½ cups cooked white rice

1 egg, lightly beaten

½ teaspoon salt substitute (page 26)

½ teaspoon freshly ground black pepper

⅛ teaspoon ground nutmeg

⅛ teaspoon ground cinnamon

8 servings

Picture: page 35

Calories	138	★★★
Fat	4g	★★★
Saturated Fat	1.4g	★★★
Cholesterol	57mg	★★★
Sodium	220mg	★★
Fiber	5.6g	★★

INGREDIENTS

8 beef boneless rib steaks (6oz each),
   trimmed of all visible fat

2 tablespoons oil

4 medium white onions, sliced
   into rings

2 tablespoons paprika

1 teaspoon cayenne pepper

2 tablespoons unbleached white flour

1 cup beef broth (page 25)

1 cup light beer

8 servings

LEMONY MEATBALL SOUP

Here is another way to extend a meat dish into a complete, healthy, balanced meal. Satisfying and warming, it makes a hearty winter dish. It is also an excellent way of using up leftover cooked meat; you can pick the last shreds off a roast to include in the meatballs. The spicing of the meatballs is particularly suited to the leek-and-lemon flavor of the soup.
Margarine usually carries the label "Not recommended for frying"; however, here you are sweating the leeks, rather than using the spread for frying, so the extra water in the spread, which might splatter during frying, will not be a problem.

· METHOD ·

Chop the leeks into 1-in rounds. Melt the margarine in a large saucepan and add the leeks. Strain the lemon juice and add it to the pot. Cover the pot tightly and cook the leeks over low heat for 10 minutes. Add the chopped celery, cover again and cook for a further 10 minutes.

    Meanwhile, combine all the meatball ingredients in a bowl and shape them into walnut-sized balls. Add the broth, parsley and meatballs to the pan and simmer for 15 minutes if the beef in the meatballs is cooked, 30 minutes if it is uncooked. Serve hot.

· SERVING SUGGESTION ·

Extra parsley or fresh coriander, chopped and sprinkled liberally over the surface, will add more vitamin C.

    Big chunks of whole wheat bread are all you need with this soup, if it is to be served as a main meal dish.

PAPRIKA BEEF

**Here is an unusual cooking method designed to reduce the need to use excess fat when cooking leaner, tougher cuts.
Boneless beef, lamb and pork steaks will taste delicious when cooked in this way.**

· METHOD ·

Place the steaks between sheets of nonstick baking paper and beat them with a mallet to flatten them and break up the fibers. Preheat the oven to 350°F.

    Heat the oil in a Dutch oven. Add the meat and onions, and sprinkle with the paprika, cayenne pepper and flour. Place the Dutch oven on high heat and shake it for 2 minutes, or until the flour turns into a starchy paste.

    Quickly pour the broth in and stir to dissolve the flour. Add the beer and cover the casserole. Transfer it to the oven and cook for 3 hours.

· SERVING SUGGESTION ·

This delicious beef stew goes well with any whole grain, but try buckwheat or millet for a new and unusual flavor. Cooking instructions will be on the package. Sprinkle the helpings with chopped fresh parsley before serving.

Calories	295	★ ★
Fat	12g	★ ★
Saturated Fat	2.7g	★ ★
Cholesterol	103mg	★ ★
Sodium	140mg	★ ★ ★
Fiber	0.8g	★

COLD SPICED BEEF

This is a very good way of pickling beef at home and is very time-saving and surprisingly easy, if you have never tackled this task before. Lean cuts of other meat – such as leg of lamb – also taste delicious when cooked in this way.
When you pickle your own beef, you know what is going into it; store-bought pickled meat contains numerous additives and preservatives required by law, which you might want to avoid.

· METHOD ·

Put the meat into a non aluminum roasting pan. Mix the peppercorns, juniper berries, pickling spice and cardamom. Sprinkle the meat with the garlic powder and then with the spice mixture, pressing it in firmly with your hands. Combine the tomato paste with the paprika, soy sauce and cider vinegar and pour it over the meat. Refrigerate for 8 hours or overnight, spooning the marinade over the meat once or twice.

Drain the meat and wrap it in foil. Roast it in the foil in a preheated 300°F oven for 1½ hours. Remove the foil carefully, reserving the meat juices. Increase the heat to 400°F and roast for 30 minutes. Let cool completely. Refrigerate covered until required. Do not slice until cold.

· SERVING SUGGESTION ·

The ideal accompaniment to pickled beef is horseradish sauce, made simply by mixing grated fresh horseradish with low-fat plain yogurt. If you cannot find fresh horseradish, you will probably be able to find a prepared horseradish in delis or supermarkets. Cold spiced beef tastes wonderful with either salad or hot vegetables. The cooking liquid can be reserved for broth or mixed with unflavored gelatin and made into aspic. This can be poured over the whole roast and left to jell, or allowed to set in a flat dish and then chopped and sprinkled over the meat slices.

INGREDIENTS

5lb boneless bottom round or
    English roast, all visible fat
    removed, tied

4 tablespoons cracked black peppercorns

6 juniper berries, crushed

2 tablespoons mixed pickling spice

½ teaspoon ground cardamom seed

1 teaspoon garlic powder

1 tablespoon tomato paste

1 teaspoon paprika

3 tablespoons soy sauce

¾ cup cider vinegar

10 servings

Picture: page 47

Calories	240	★ ★ ★
Fat	9g	★ ★
Saturated Fat	3.5g	★ ★
Cholesterol	112mg	★ ★
Sodium	120mg	★ ★ ★
Fiber	0.0g	★

LEAN TERRINE

for recipe see page 31

LEMONY MEATBALL SOUP

for recipe see page 32

INGREDIENTS

8 large whole green cabbage leaves

1lb lean ground beef

¼ teaspoon salt

½ teaspoon ground cumin

Juice and grated rind of 2 lemons

*1 cup blanched almonds, slivered
    or coarsely chopped*

1 onion, grated

1¾ cups cooked short-grain brown rice

½ teaspoon black pepper

4 tablespoons finely chopped parsley

2 cups tomato juice

1 teaspoon paprika

½ teaspoon cayenne pepper

⅓ cup golden raisins

4 servings

Picture: page 38

Calories	447	★ ★
Fat	19g	★
Saturated Fat	3.2g	★ ★
Cholesterol	66mg	★ ★ ★
Sodium	676mg	★
Fiber	11.0g	★ ★ ★

CABBAGE LEAVES WITH BEEF AND ALMOND STUFFING

**Meat-stuffed vegetables are a healthy combination.
They make a little meat go a long way and, when
made with lean meat, are low in fat.
The meat is doubly tenderized, since it is both ground and
slow-cooked. A huge variety of vegetables can be stuffed, and there
are endless delicious combinations of meat stuffings. In the Middle
East, cooks will spend all day filling tiny vegetables, such as tiny
zucchini or even okra! Cabbage leaves are more convenient for the
modern cook, since their preparation is quick and easy.
You can use either white or brown rice in the recipe, but it should be
short-grained, so that it will stick together.**

· METHOD ·

Make sure the cabbage leaves are free of blemishes or holes. Plunge them into boiling water for 3 minutes to soften the veins, so that they can be rolled easily. Leave them to drain.

Put the beef into a dry nonstick skillet and cook, stirring frequently over medium heat to break up any lumps, until the fat begins to run out. Sprinkle the meat with the salt, cumin and the grated lemon rind. Add the almonds and grated onion, with any onion juice, and stir again.

Remove the meat mixture from the skillet and transfer it to a bowl. Mix it well with the rice, pepper and parsley, until it coheres into a ball. Divide it into 8 equal portions.

Lay out a cabbage leaf on a work surface. Place the stuffing in the center, then fold the bottom end of the leaf toward the center. Do the same with the two sides, then roll the leaf up like a jelly roll, with the ends tucked securely inside. Tie the rolls with strong, thin thread. Pour the tomato juice into a large ovenproof casserole with a lid. Add the lemon juice, paprika, cayenne and golden raisins. Arrange the cabbage rolls neatly in the sauce. Cover the casserole and cook in a preheated 350°F oven for 20 minutes, then turn the cabbage rolls over carefully, and bake them for another 10 minutes. If the sauce is too runny for your taste, cook the cabbage leaves uncovered for the last 10 minutes. Remove the threads tying the rolls before serving. Serve hot.

· SERVING SUGGESTION ·

Plain low-fat yogurt provides a delicious counterbalance to the sweet-and-sour flavor of the cabbage rolls. These servings are very generous and constitute a meal in themselves, but if another starchy vegetable is required, serve the cabbage rolls with sweet corn kernels, which make an appealing color contrast.

BEEF NAPOLEON

If Beef Wellington is fillet of beef baked in a pastry crust, then Beef Napoleon is the same dish, baked in a yeast dough crust. The effect is just the same, but most of the fat normally used in the preparation of the former classic dish has been left out, and the result is just as delicious. It is an expensive dish for special occasions.

· METHOD ·

Heat the oil in a nonstick skillet. Brown the meat all over on medium heat. Remove it and add the shallots. Cook on low heat for 3 minutes, add the mushrooms and season with the pepper and chopped parsley. Cook until the mushrooms are soft, stirring occasionally. Add the port and cook for 3 more minutes, or until the skillet is dry.

Flour a large cookie sheet, and sprinkle it with semolina. Roll out the dough on a well-floured surface into a rectangle large enough to wrap up the meat. Brush it with the egg white to seal, and spread it with the mushroom mixture, leaving a border of at least 2in of dough. Place the fillet in the center of the dough.

Fold the short sides of the rectangle toward the center, then fold in the long sides toward the center, overlapping them. Brush all the seams with water to make them stick, and pinch them lightly to seal. Carefully transfer the fillet to the prepared cookie sheet. Let stand for 30 minutes to proof the dough.

Heat the oven to 400°F. Check that the seams are still tightly sealed and, if not, pinch them closed again. Turn the dough parcel over so that the long seam is on the bottom.

Brush the dough all over with the milk. Lift the parcel slightly to get underneath it; then pierce three holes in the top with a metal skewer, to allow steam to escape during baking. Put the beef into the oven and bake for 20 minutes; reduce the heat to 275°F and bake for another 20 minutes. Slide the beef onto a warmed serving dish. Keep it warm and do not slice it until it comes to the table.

· SERVING SUGGESTION ·

Fresh green beans, asparagus tips and other tiny vegetables are fitting accompaniments for this elegant dish. No extra starch is needed, since it is provided by the bread dough. If the rectangle can be made without using up all the dough, extra scraps can be rolled and cut into decorative shapes. Place them in an attractive arrangement on top of the dough before baking.

INGREDIENTS

1 tablespoon oil

2½lb beef rib eye roast, in one piece

3 shallots, chopped

3 cups button mushrooms, thinly sliced

½ teaspoon black pepper

2 tablespoons chopped parsley

2 tablespoons port

Flour and semolina for sprinkling

½ recipe white bread dough (page 26)

1 egg white, lightly beaten

3 tablespoons skimmed milk

8 servings

Calories	411	★ ★
Fat	9g	★ ★
Saturated Fat	2.6g	★ ★
Cholesterol	74mg	★ ★ ★
Sodium	214mg	★ ★
Fiber	3.4g	★ ★

CABBAGE LEAVES WITH BEEF AND ALMOND STUFFING
for recipe see page 36

INGREDIENTS

1½ cups whole wheat berries

1 tablespoon oil

3 onions, sliced

1 teaspoon salt

1 teaspoon black pepper

1 teaspoon ground allspice

1 teaspoon chili powder

1½ lb lean stewing beef, cut into small
    cubes, trimmed of all visible fat

3 tomatoes, peeled and seeded

4 tablespoons tomato paste

6 servings

Picture: page 51

Calories	341	★ ★
Fat	10g	★ ★
Saturated Fat	3.1g	★ ★ ★
Cholesterol	79mg	★ ★
Sodium	290mg	★ ★
Fiber	2.8g	★

INGREDIENTS

1½ lb lean ground beef

1 medium onion, grated

1 large garlic clove, finely chopped

¼ cup whole wheat breadcrumbs

¼ cup grated Parmesan cheese

4oz pine nuts (pignolas)

6 tablespoons chopped parsley

1 egg

1 teaspoon cayenne pepper

6 servings

Picture: page 55

Calories	334	★ ★
Fat	20g	★
Saturated Fat	4.5g	★ ★
Cholesterol	123mg	★ ★
Sodium	202mg	★ ★
Fiber	1.4g	★

CLAY-COOKED BEEF-WHEAT STEW

This delicious stew can be cooked in an ordinary casserole if 2 cups broth is added to the pot, and the stew is put into a preheated oven.

· METHOD ·

Soak an unglazed clay pot in cold water, and soak the wheat in water to cover for 2 hours. Heat the oil and fry the onions, stirring frequently, until they are transparent, then sprinkle them with the spices. Add the meat to the pan and cook it, stirring, until it is no longer red on the outside.

    Drain the clay pot and wipe the outside to remove excess moisture. Transfer the meat mixture to the pot and add the wheat, tomatoes and tomato paste. Place in a cold oven and turn the heat to 350°F. Bake for 2 hours, checking after 90 minutes to see if the wheat is drying out. If necessary, add 6 tablespoons hot water to the pot. Serve hot.

· SERVING SUGGESTION ·

A salad of romaine Lettuce, lemon juice and lots of chopped fresh parsley, tossed with a few 1-in squares of toasted pita, makes a delicious accompaniment.

BROILED PATTIES WITH PARMESAN AND PINE NUTS

**The leaner the beef, the better it is for broiling, and cheaper cuts, like skirt steak, make very good ground meat.
Always buy whole pieces of meat and get the butcher to grind it for you, or grind it yourself at home; the ready-ground meat will be fattier and not as fresh. This recipe extends and flavors the meat, avoiding the need for salt.**

· METHOD ·

Combine all the ingredients in a large bowl. Shape into rounds about 2½ in thick. Broil over a barbecue or under a hot broiler until the outsides are crisp and browned, about 3 to 5 minutes on each side, depending on how rare you like them. Eat while very hot.

· SERVING SUGGESTION ·

Baked potatoes, moistened with low-fat yogurt or margarine, would make good accompaniments to these patties. A green salad, with a few leaves of radicchio and chicory included with the lettuce, would round off the meal.

TOURNEDOS WITH MOZZARELLA AND GREEN SALAD

This meal is simplicity itself but elegant enough to be served at the most important dinner party. However, it is rather rich, so it should only be eaten as an occasional treat!
The tournedos will not need salt, as the cheese provides sufficient flavor. Instead of mozzarella slices, an interesting substitute is ricotta or Swiss cheese, or even Muenster.
Many people love garlic, but if you find it indigestible or are afraid your guests might, cut out the green shoot at the base of each clove before use; this often solves the problem.

· METHOD ·

Rub the steaks with the garlic cloves, and rub any remaining garlic around the salad bowl. Then make the salad and dressing. Combine the cabbage and greens in a bowl, stirring well. Mix the dressing ingredients, preferably in a blender or food processor, and pour the dressing into a jar. Cover the salad and jar, and refrigerate until required.

Heat the broiler to maximum. Place the tournedos under it, about 2in from the heat. Broil the steaks for 2 minutes to seal the juices. Turn them over and broil the other side for 3 to 5 minutes, depending on how rare you want them. Trim away all visible fat from the steaks. Slice the cheese crosswise into 4 pieces. Place each steak, well-broiled side upward, on a cheese slice, and trim around the cheese so that it will fit neatly over the steak during cooking.

Place the steaks back onto the broiler pan, cheese-covered side uppermost. Sprinkle with the pepper and broil on high for 3 minutes. Serve immediately with the salad and serve the dressing separately.

· SERVING SUGGESTION ·

Potatoes baked in their jackets are the obvious accompaniment to a grill or barbecue. Sweet potatoes, parboiled, then baked in their jackets for 30 minutes or puréed with a little margarine, make a more unusual side dish.

INGREDIENTS

4 tournedos steaks (about 6oz each) brought to room temperature and trimmed of all visible fat
4 garlic cloves
2½ cups red cabbage, cut into strips
1 small head Boston lettuce, cut into strips
1 head arugula, leaves separated
1 cup spinach leaves, cut into strips
8oz mozzarella cheese
½ teaspoon black pepper

SALAD DRESSING

½ cup plain yogurt
2 garlic cloves, finely chopped
2 tablespoons finely chopped fresh coriander
1 cucumber, peeled and finely chopped

4 servings
Picture: page 43

Calories	272	★ ★
Fat	16g	★
Saturated Fat	9.7g	★
Cholesterol	73mg	★ ★ ★
Sodium	317mg	★ ★
Fiber	3.7g	★ ★

DRY-FRIED BEEF WITH ORANGE BITTERS

for recipe see page 44

TOURNEDOS WITH MOZZARELLA AND GREEN SALAD
for recipe see page 41

INGREDIENTS

1-lb piece porterhouse or flank steak, trimmed of fat

2 tablespoons oil

2 tablespoons dry sherry

2 tablespoons orange bitters

1 tablespoon hoisin sauce

1 garlic clove, finely chopped

½ teaspoon salt

2 medium carrots, grated

2 green onions, finely chopped

1-in piece ginger root, finely chopped

2 seedless tangerines, peeled, segments separated

½ teaspoon black pepper

4 servings

Picture: page 42

Calories	234	★ ★ ★
Fat	13g	★ ★
Saturated Fat	3.1g	★ ★
Cholesterol	66mg	★ ★ ★
Sodium	330mg	★ ★
Fiber	1.2g	★

INGREDIENTS

1lb rump steak or top round, frozen

2 large onions

3 garlic cloves, finely chopped

10 black peppercorns, crushed

2 tablespoons soy sauce

2 tablespoons honey

1 tablespoon oil

1 large tomato, skinned and chopped

¼ cup water

1 stick cinnamon

2 whole cloves

4 servings

DRY-FRIED BEEF WITH ORANGE BITTERS

**Dry-frying is a Chinese technique typical of Szechwan province. It was developed to compensate for lack of fuel.
This is yet another Oriental technique developed to cook food as quickly as possible. It has the added advantage of also being a low-fat cooking technique and thus particularly suitable for adaptation to modern requirements.
In Szechwan, this type of dish would include very hot spices, such as chili oil; this adaptation for Western palates has more subtle flavoring.**

· METHOD ·

Use a very sharp knife to cut the beef into thin slices about 1in wide. Heat a wok or nonstick skillet over high heat. Add the oil. When it is hot, add the beef and the sherry, stirring until the beef separates into shreds.

Reduce the heat and drain off any liquid remaining in the pan. Stir gently until the beef is dry. Add the hoisin sauce, garlic, salt, the rest of the sherry and the bitters. Stir a few times and increase the heat to high, and add the carrots. Stir-fry for a minute, then add the onions, ginger, tangerines and pepper. Stir-fry for a further 2 minutes. Serve immediately.

· SERVING SUGGESTION ·

This beef dish would go well with a slimming coleslaw – raw shredded white cabbage with a little grated carrot, bound with low-fat yogurt or Special Low-Fat Mayonnaise (page 26) – or with braised or steamed celery. Add brown rice or brown noodles for the perfect meal.

BEEF JAVA

**Marinating meat and beating it to break up the fibers are both tenderizing techniques that make it more digestible.
Meat stir-fried in the Oriental way cooks quickly, thus sealing in the juices and reducing the amount of oil absorbed during cooking.
However, such quick cooking can make meat tough and leathery if it is not pretenderized.**

· METHOD ·

To enable the beef to be sliced as thinly as possible, freeze it first. Slice it in thin strips and trim all visible fat from the meat. Allow the meat to thaw, then lay the slices between two sheets of plastic or nonstick baking paper and beat them with a meat pounder to flatten them.

Mince one of the onions and combine with the chopped garlic. Add the peppercorns, soy sauce and honey and mix well. Put the meat into a shallow bowl, pour the marinade over it and stir to coat well. Leave to marinate for at least 1 hour, turning every 15 minutes. Alternatively, it can be marinated in the refrigerator for several hours or overnight, but remember to turn it at least four times during the marinating.

Slice the remaining onion thinly, lengthwise. Heat the oil in a wok or skillet and add the sliced meat and the marinade. Stir-fry for 2 minutes, then add the tomato, water, cinnamon and cloves. Cook, stirring, until the meat is tender and most of the liquid has evaporated, about 7 minutes. Discard the cinnamon and cloves before serving.

· SERVING SUGGESTION ·

Serve with a salad containing mixed pickled vegetables, or with sauerkraut. Accompany this with brown rice or noodles, or Japanese buckwheat noodles (soba).

Calories	242	★ ★ ★
Fat	10g	★ ★
Saturated Fat	2.5g	★ ★
Cholesterol	74mg	★ ★ ★
Sodium	89mg	★ ★ ★
Fiber	2.3g	★

BARBECUED RIBS WITH ORANGE-MUSTARD SAUCE

Although this recipe uses beef short ribs, veal ribs or pork spareribs can also be used.
Ribs can be fatty, so be sure to remove every trace of surplus fat before cooking. Barbecuing is the best cooking method for quickly melting away as much fat as possible.
This recipe would make a lovely summer children's party treat, and the blackstrap molasses is an excellent source of iron, too.

INGREDIENTS

1 cup diabetic or low-sugar orange marmalade

6 tablespoons cider vinegar

¼ cup blackstrap molasses

2 tablespoons honey

4lb beef short ribs, trimmed of excess fat

6 servings

· METHOD ·

Combine the marmalade, vinegar, blackstrap molasses and honey in a large, deep bowl. Add the ribs and stir well. Marinate the mixture for several hours or overnight, stirring occasionally.

To cook, remove the ribs from the marinade and broil them on a barbecue or under a very hot broiler for 15 minutes, turning frequently.

Meanwhile, warm the marinade. Serve the ribs hot, with the marinade as a sauce.

· SERVING SUGGESTION ·

A green salad is always appropriate with such a substantial dish, and potatoes baked or boiled in their jackets are also an obvious choice. If new potatoes are in season, parboil them for 15 minutes, then thread them, unpeeled, on skewers and brush them with oil. Roast them over the hot coals for 5 minutes, turning frequently. Serve with low-fat plain yogurt, sprinkled with chopped chives or green onions.

Calories	268	★ ★
Fat	6g	★ ★ ★
Saturated Fat	2.3g	★ ★
Cholesterol	74mg	★ ★ ★
Sodium	99mg	★ ★ ★
Fiber	0.2g	★

PERSIAN MEATBALLS WITH SPINACH

for recipe see page 48

COLD SPICED BEEF

for recipe see page 33

INGREDIENTS

1lb lean ground beef

1 medium onion, grated

2 slices whole wheat bread, crusts removed, soaked in water and squeezed dry

½ teaspoon salt

1 teaspoon ground coriander

1 teaspoon ground cumin

½ teaspoon ground black pepper

½ teaspoon ground cinnamon

2 tablespoons oil

2 garlic cloves, crushed

5 cups finely chopped fresh spinach

6 tablespoons chopped parsley

Juice of 4 oranges

Juice of 2 lemons

1 tablespoon cornstarch

6 servings

Picture: page 46

Calories	227	★ ★ ★
Fat	11g	★ ★
Saturated Fat	2.4g	★ ★
Cholesterol	44mg	★ ★ ★
Sodium	283mg	★ ★
Fiber	2.1g	★

PERSIAN MEATBALLS WITH SPINACH

In regions such as the Middle East, where meat is lean but tough and expensive, it is very often ground and added to dishes in small proportions. This is a practice we in the West would do well to adopt, as it turns a starch-and-vegetable dish into a complete, healthy, balanced meal. Adding soaked and squeezed bread to ground meat – a trick used in *haute cuisine* to make the mixture light and fluffy – is not merely economical; it is a positive step toward healthier eating.

· METHOD ·

Combine the meat, onion, squeezed bread, salt and spices in a large bowl. When the mixture is smooth, shape it into balls the size of a walnut.

Heat 1 tablespoon of the oil in a nonstick skillet with a tight-fitting lid. Add the meatballs and sauté them, turning frequently until they are evenly browned, about 7 minutes. Remove the meatballs and drain them on paper towels.

Heat the rest of the oil in the skillet and add the garlic, spinach and parsley. Cook, stirring and tossing, until the spinach has wilted, about 3 minutes. Return the meatballs to the pan and add 1 cup water. Cover the pan and simmer on very low heat for 15 minutes.

Combine the juices of the orange and lemon with the cornstarch. Add this mixture to the pan and stir well to combine. Serve immediately.

· SERVING SUGGESTION ·

These meatballs have their own vegetables cooked along with them, but why not try serving a raw spinach salad as well? This salad consists of well-washed spinach, finely shredded and dressed with lemon juice, black pepper and bacon-flavored soy bits. Brown rice, sautéed in a tablespoon of oil, then steamed in a tightly closed pot for 45 minutes with a cloth on top to absorb the steam, is an alternative Persian-style accompaniment.

SWEET-AND-SOUR BEEF SALAD

**Corned beef and smoked beef are both very popular but, like all
salty foods, they should not be consumed too often.
This salad combines cooked and raw ingredients, and the potatoes
turn it into a complete meal.**

· METHOD ·

Bring 4 cups water to the boil in a saucepan. Clean and trim the potatoes,
carrots, onions and beans. Add the potatoes, carrots and onions, and
parboil for 15 minutes. Drain and cool (saving the cooking water to use for
broth). Slice the onions crosswise into rings.

Slice the mushrooms, and sprinkle them with 1 tablespoon of the cider
vinegar. Core and slice the apple but do not peel it – sprinkle it with the
rest of the cider vinegar. Slice the pickles into rounds. Combine these with
the cooked vegetables. Mix the dressing ingredients, and pour the liquid
over the vegetables, stirring well.

Slice the beef lengthwise, and arrange the strips around the edge of a
large plate. Pile the cooked vegetables in the center. Chop the parsley and
sprinkle it liberally over the salad before serving. Garnish with strips of
tomato or red pepper.

· SERVING SUGGESTION ·

Serve with extra pickled vegetables, such as mustard or sweet pickles and
raw bean sprouts or alfalfa sprouts, plus a potato salad made with Special
Low-Fat Mayonnaise (page 26). This dish can be prepared well in advance
and is ideal for guests, especially if they are coming from far afield and
cannot be relied upon to turn up at an exact time.

INGREDIENTS

*1lb sliced, cooked, corned beef or
    salted brisket, silverside, or
    pastrami*
8oz new potatoes
8oz baby carrots
4 small onions, peeled
2 cups/8oz sliced green beans
1 red bell pepper, seeded and sliced
1 cup button mushrooms
1 large tart apple
2 tablespoons wine or cider vinegar
4 pickles
1 small bunch parsley
*strips of tomato or red pepper, for
    garnish*

FOR THE DRESSING

1 tablespoon olive oil
1 teaspoon honey
4 tablespoons red wine or cider vinegar
1 teaspoon Worcestershire sauce
1 garlic clove, finely chopped
½ teaspoon black pepper

6 servings

Calories	239	★ ★ ★
Fat	7g	★ ★ ★
Saturated Fat	2.1g	★ ★
Cholesterol	62mg	★ ★ ★
Sodium	828mg	★
Fiber	6.1g	★ ★

BEEF TERIYAKI
for recipe see page 52

CLAY-COOKED BEEF-WHEAT STEW
for recipe see page 40

INGREDIENTS

1½lb lean beef, ground

½ cup uncooked short-grain
    brown rice

⅓ cup seedless raisins

⅓ cup currants

½ cup blanched, slivered almonds

¼ teaspoon black pepper

¼ teaspoon ground turmeric

¼ teaspoon ground cinnamon

Juice and grated rind of 2 lemons

1 medium pumpkin (about 5lb)

2 tablespoons brown sugar

½ teaspoon salt

4 servings

Calories	528	★
Fat	15g	★ ★
Saturated Fat	3.7g	★ ★
Cholesterol	100mg	★ ★
Sodium	428mg	★
Fiber	7.1g	★ ★

INGREDIENTS

¼ cup light soy sauce

¼ cup sweet rice wine or sherry

2 garlic cloves, finely chopped

1-in piece ginger root, grated

6 thin beef steaks (6oz each),
    trimmed of all visible fat

1 tablespoon oil

2¼ cups bean sprouts

⅓ cup alfalfa sprouts

6 servings

Picture: page 50

WHOLE STUFFED PUMPKIN

Although life may be too short to stuff a mushroom, it's not too short to stuff a pumpkin. In a recent survey, people named pumpkin as their least favorite vegetable, but they had never tasted it cooked like this.
It's an option worth choosing, because vegetables of the pumpkin family consist mainly of water and thus are low in calories. Any large winter squash can be treated in the same way.

· METHOD ·

Put the meat into a nonstick skillet and cook it, stirring constantly to break up any pieces that stick together, until it is no longer red. Throw the rice into briskly boiling water, parboil it for 20 minutes and drain it thoroughly. Mix it with the raisins, currants, almonds, pepper, turmeric and cinnamon. Combine thoroughly with the meat. Sprinkle with the lemon juice.

Cut the top off the pumpkin about 1 in below the stem and reserve this "lid". Remove the seeds and pith with a large spoon and discard them. Scoop out some of the flesh, being careful not to pierce the skin, until there is a good-sized cavity.

Preheat the oven to 350°F. Stuff the pumpkin with the meat mixture and cover it with the "lid". Wrap the whole pumpkin in foil and place it on a baking sheet. Bake for 1½ hours. Leave the foil to cool for a few minutes, then remove it carefully to avoid being scalded by escaping steam. To serve, stand the pumpkin on its side and slice it into rings, so that each guest gets some filling inside a pumpkin ring.

· SERVING SUGGESTION ·

This is a completely balanced meal in itself, but a cooked green vegetable, such as cabbage, brussels sprouts or spinach, would go well with it.

BEEF TERIYAKI

Although some Japanese dishes are complicated to prepare and require unusual ingredients, Beef Teriyaki is extremely simple. It is typical of the economical use of meat by the Japanese and their dislike of excessively fatty foods.

· METHOD ·

Mix the soy sauce, sweet rice wine or sherry, garlic and ginger root together in a shallow dish. Add the steaks. Marinate them in the mixture for at least 2 hours at room temperature.

Heat the oil in a nonstick skillet or wok. Drain the steaks and add them to the skillet. Fry quickly until they are no longer red. Pour the marinade over and continue cooking for 5 minutes, stirring frequently, until the meat looks glazed.

Remove it from the pan and reserve it on a serving dish. Toss the bean sprouts and alfalfa sprouts into the pan and stir-fry for 2 minutes. Garnish the meat with the vegetables.

· SERVING SUGGESTION ·

The traditional accompaniment for this dish is Japanese short-grained white rice. You can try brown rice – perhaps with a few raisins added – or brown noodles.

Calories	279	★ ★
Fat	11g	★ ★
Saturated Fat	3.7g	★ ★
Cholesterol	103mg	★ ★
Sodium	111mg	★ ★ ★
Fiber	0.3g	★

STIR-FRIED BEEF WITH PEPPERS AND SNOW PEAS

Stir-frying, the quick-frying method used in the Orient, is a low-fat method of frying meat, even though extra oil is often required. The oil used should always be high in polyunsaturates. Grape seed oil, imported from France, which is just becoming widely available, is ideal, but if you cannot get it use safflower oil.

· METHOD ·

Use a very sharp knife to slice the beef as thinly as possible. Cut it into strips about 1in wide. Combine the salt substitute, soy sauce, dry sherry, honey and cornstarch in a bowl. Add the beef and stir well.

While the beef is marinating, prepare the vegetables. Seed the peppers and chilies. Cut the peppers into small dice, and chop the chilies. Trim and chop the green onions and trim the snow peas. Peel the ginger root and grate it, or chop it finely.

Heat a wok or large skillet over high heat. Add the oil; when it starts to smoke, add the beef and stir-fry for a minute. Remove the beef with a slotted spoon.

Add the prepared vegetables and the bean sprouts to the skillet, and stir-fry for 2 minutes. Return the beef to the pan and stir-fry for a further 2 minutes. Serve immediately.

· SERVING SUGGESTION ·

Of course, the classic accompaniment is rice, but wholewheat noodles are quick to prepare and are equally good with this dish. Serve a tomato salad as a side dish; it will make an attractive color contrast.

INGREDIENTS

10oz lean beef (flank or rib eye)
¼ teaspoon salt substitute (page 26)
1 tablespoon soy sauce
1 tablespoon dry sherry
1 teaspoon honey
1 tablespoon cornstarch
2 large green bell peppers
2 small green chilies (such as jalapeños)
3 green onions
2 cups snow peas
½in ginger root
3 tablespoons oil
2¼ cups bean sprouts

4 servings
Picture: page 54

Calories	293	★ ★
Fat	15g	★ ★
Saturated Fat	2.9g	★ ★
Cholesterol	41mg	★ ★ ★
Sodium	52mg	★ ★ ★
Fiber	5.1g	★ ★

STIR-FRIED BEEF WITH PEPPERS AND SNOW PEAS

for recipe see page 53

BROILED PATTIES WITH PARMESAN AND PINE NUTS
for recipe see page 40

VEAL

Choose veal that is pale pink and juicy-looking when raw, and avoid meat that is flabby and brightly colored. It should be eaten within 24 hours of purchase.

Veal is tougher than beef because it contains less fat and more connective tissue, so it needs very thorough cooking. It is rarely broiled because the meat fibers are too tough for this treatment. Veal also benefits particularly from being beaten with a meat pounder to soften the fibers. USDA gradings for veal are the same as for beef, and the same rules apply to choosing the grade of meat right for you.

You will need about 6oz meat per person for boneless veal and 8oz per person for veal chops and cutlets.

EGG-AND-VEAL ROLLS

This is a low-fat version of a delicious Italian dish from the region of Mantua. It is both attractive and nutritious and is always eaten cold. The egg is used as a "stretcher" to make the meat go further. If you have a microwave oven, you can reduce the fat content even more by cooking the omelets without any fat.
If each egg-and-pea omelet is placed in a round shallow dish and microwaved on high for 2 minutes, shutting off the power for a second or two after 1 minute, it will be cooked through and will taste just as good as if it had been shallow-fried.

INGREDIENTS

4 lean veal scallops (about 6oz each)
½ tablespoon light cooking oil
4 eggs
1 cup cooked green peas
¼ teaspoon salt
¼ teaspoon freshly ground black pepper
½ teaspoon dried sage
1 bay leaf
1 cup chicken or beef broth
   (page 24)
2 tablespoons chopped parsley

4 servings
Picture: page 58

· METHOD ·

First prepare the scallops by putting them between 2 sheets of nonstick baking paper. Beat them with a meat pounder to a very thin, even thickness, as you would for Wiener Schnitzel.

Heat the oil in a skillet. Beat one of the eggs and stir in a quarter of the peas. Pour the mixture into the pan and cook over medium heat on one side for 2 minutes. Flip over and cook for 1 minute on the other side. Remove this omelet from the skillet and drain it on paper towels. Repeat with the rest of the eggs and peas, until you have 4 omelets.

Remove the paper from the veal and lay an omelet over each scallop. Season with salt and pepper. Roll up each scallop like a jelly roll, and secure it with a wooden toothpick.

Put the sage and bay leaf into a pot with a tight-fitting lid. Add the veal rolls and the broth. Cover and bring to the boil, then reduce the heat and simmer until the veal is tender, about 1 hour. Leave the rolls to cool in the liquid until they are at room temperature.

Remove the rolls and drain them. Slice them across into ¼-in slices with a sharp knife and arrange them in a serving dish. Sprinkle them with the chopped parsley before serving.

· SERVING SUGGESTION ·

Serve the cold veal with a Russian salad (a mixture of diced vegetables and peas), bound with Special Low-Fat Mayonnaise (page 26), and a green salad. The cold cooking liquid can be thickened with low-fat yogurt and served as a sauce.

Calories	343	★★
Fat	14g	★★
Saturated Fat	4.6g	★★
Cholesterol	428mg	★
Sodium	449mg	★
Fiber	5.9g	★★

EGG AND VEAL ROLLS

for recipe see page 57

VEAL SCALLOPS WITH SAGE AND LEMON
for recipe see page 61

INGREDIENTS

*2lb boned loin or boned and rolled
    shoulder of veal*

⅔ cup white wine

1 lemon, sliced

1 seedless orange, sliced

1 onion, chopped

1 teaspoon ground coriander

6 tablespoons chopped fresh coriander

6 servings

Calories	213	★ ★ ★
Fat	5g	★ ★ ★
Saturated Fat	1.4g	★ ★ ★
Cholesterol	150mg	★ ★
Sodium	193mg	★ ★ ★
Fiber	1.0g	★

INGREDIENTS

2 lemons

*8 veal cutlets, trimmed of all surplus
    fat (3oz each)*

½ teaspoon salt

½ teaspoon white pepper

*4oz uncooked tagliatelle verdi
    (green noodles)*

1 cup tomato juice

½ teaspoon basil

2 tablespoons chopped parsley

4 servings

DRIP-ROAST VEAL WITH CORIANDER

**Although veal is from a young animal, the meat must be well
cooked right through, unlike beef, or it becomes tough.
This drip-roasting method is suitable for all roasts, especially those
that have a generous amount of fat.**

· METHOD ·

Put the meat into a bowl. Pour the wine over it and sprinkle with the
lemon, orange, onion and ground coriander. Leave to marinate for 3 hours
at room temperature.

Heat the oven to 350°F. Drain the veal, reserving the marinade, and
place on a trivet in a roasting pan. Sprinkle it with half the fresh coriander.
Roast it for 90 minutes, basting with the marinade every 15 minutes.

Remove the meat and roasting pan from the oven. Lift the roast onto a
plate and let it rest for 10 minutes before slicing. Meanwhile, put the pan
over high heat and stir to dislodge any bits that have stuck to the bottom.
When the liquid boils, strain it into a bowl. Leave it to cool, then
refrigerate. When the surface fat has congealed, remove it with a knife.
The liquid can be used as gravy for another dish.

Slice the veal and arrange attractively on a serving plate. Sprinkle the
slices with the rest of the fresh coriander.

· SERVING SUGGESTION ·

Any remaining marinade can be heated and served as a sauce. Stewed
mushrooms and spinach make excellent accompaniments. Potatoes in
their jackets can be baked with the veal; serve them with yogurt and
chopped chives.

VEAL PICCATA WITH TAGLIATELLE VERDI

**This dish is simplicity itself, and the veal flavor is
not masked by other flavorings.
Normally, veal piccata is full of oil and butter, but this is a low-fat
version, which is just as tasty as the original.
Serve very hot.**

· METHOD ·

Grate the rind from one of the lemons, and squeeze the juice. Slice the
other lemon crosswise into the thinnest possible slices. Heat a nonstick
skillet. Put the cutlets into the pan and sprinkle with the salt, pepper and
lemon rind. Cook just until they are no longer pink, about 2 minutes on
each side. Reserve in a warm place.

Bring a pot of water to the boil. Add the noodles and cook for 12
minutes or as recommended on the package.

In a small saucepan, heat the tomato juice and add the basil. Drain the noodles. Divide them into four "nests" and arrange them on four plates. Return the veal to the pan and brown on one side. Add the lemon slices to the pan and brown the veal on the other side. Arrange a cutlet on each noodle "nest", arrange the lemon slices on top of them, and sprinkle with parsley. Pour a little pool of the tomato sauce onto each plate. Serve hot.

· SERVING SUGGESTION ·

A tomato salad, consisting of sliced tomatoes tossed with fresh herbs of your choice and moistened with a few drops of oil low in polyunsaturates, is the classic accompaniment to this dish. Whole wheat bread sticks (grissini) will add an Italian touch.

Calories	269	★ ★
Fat	5g	★ ★ ★
Saturated Fat	1.7g	★ ★ ★
Cholesterol	135mg	★ ★
Sodium	414mg	★
Fiber	1.3g	★

VEAL SCALLOPS WITH SAGE AND LEMON

Here is another example of how sharp flavors – in this case lemon juice combined with herbs or spices – can provide a natural salt substitute. Furthermore, the full flavor of the veal comes through and is not marred by heavy, greasy coatings.
If you cannot get fresh sage leaves, use two teaspoons of dried sage.

INGREDIENTS

1 garlic clove, finely chopped
1 tablespoon oil
Juice of 2 lemons
½ teaspoon black pepper
2 tablespoons fresh sage leaves
4 veal scallops (5oz each), trimmed of all fat

4 servings
Picture: page 59

· METHOD ·

Combine the garlic, oil, lemon juice, pepper and half the sage leaves in a bowl. Add the veal and marinate for 30 minutes, turning occasionally. Heat the broiler and place the veal under it. Broil for 4 minutes on each side, or until well browned. Sprinkle with the rest of the sage leaves before serving.

· SERVING SUGGESTION ·

A salad of cooked vegetables – such as a Russian salad in Special Low-Fat Mayonnaise (page 26), or cold, cooked green beans – would go well with this veal dish. For a starchy accompaniment, serve the scallops on whole wheat toast.

Calories	201	★ ★ ★
Fat	8g	★ ★ ★
Saturated Fat	2.1g	★ ★
Cholesterol	135mg	★ ★
Sodium	165mg	★ ★ ★
Fiber	0.0g	★

VEAL STEAKS WITH RATATOUILLE
for recipe see page 65

MEDALLIONS OF VEAL WITH SPRING VEGETABLES
for recipe see page 64

INGREDIENTS

1 tablespoon diet margarine

8oz pearl onions, peeled

3 garlic cloves, crushed

8oz leeks, white parts only, sliced into 1-in pieces

8oz baby carrots, trimmed and scraped

1 celery stalk

1 bay leaf

½ teaspoon dried thyme

2 tablespoons chopped parsley

2 teaspoons fresh marjoram or thyme, finely chopped

1 medium Boston or Bibb lettuce, coarsely shredded

Water or broth (optional)

1lb zucchini, sliced into 1-in rounds

1lb snow peas, trimmed

6 veal medallions (about 6oz each)

Black pepper

Juice and grated rind of 1 lemon

6 servings

Picture: page 63

Calories	362	★ ★
Fat	8g	★ ★
Saturated Fat	3.0g	★ ★
Cholesterol	210mg	★
Sodium	355mg	★ ★
Fiber	9.2g	★ ★ ★

MEDALLIONS OF VEAL WITH SPRING VEGETABLES

**Medallions are a lean cut from the rump end of the fillet, the veal equivalent of the tournedos in beef.
This cut is expensive but low in fat. It is usually sold with a piece of fat wrapped round it, which should be discarded.**

· METHOD ·

Melt the margarine in a large flameproof casserole. Add all the vegetables, except the zucchini and snow peas, in the order in which they are listed, sprinkling the shredded lettuce over the top. Cover the casserole tightly and simmer on a very low heat for 30 minutes, checking at least twice to see that there is still liquid in the bottom of the pan and that the vegetables are not drying up or browning. If they are, add a couple of tablespoons of water or broth. Add the zucchini and snow peas after 30 minutes.

Heat the broiler and put the veal medallions under it. Sear for 2 minutes to seal in the juices, then turn over and cook for 5 minutes. Turn again and cook for 3 to 5 minutes, depending on how well done you like them. Sprinkle with black pepper and a squeeze of lemon juice, as well as the grated lemon rind.

The vegetables should now be cooked. Sprinkle them with the rest of the lemon juice, and serve immediately with the veal.

· SERVING SUGGESTION ·

This dish is complete in itself. However, if you feel the need for a starchy vegetable, then the obvious choice is new potatoes – steamed or boiled – garnished with yogurt or low-fat margarine and a few fresh mint leaves, left whole.

VEAL STEAKS WITH RATATOUILLE

Veal steaks are large lean cuts, with a ring of fat around them which should be removed before cooking. The combination of veal and ratatouille is quite a common one, and very delicious. Veal is more easily digestible than beef, as the fibers are softer.

· METHOD ·

In a large nonstick skillet with a lid, sauté the veal steaks for 3 minutes on one side to seal the juices, then turn over and sauté on the other for another 3 minutes. Remove the steaks from the skillet, wipe it with a paper towel and add the olive oil. Heat the oil and add the onions. Cook, stirring, until the onions are golden, about 10 minutes. Then add the tomatoes, eggplant, zucchini, red peppers and garlic. Return the steaks to the pan and add the sage, bay leaf and lemon peel. Cover and simmer for 1 hour on very low heat.

Discard the bay leaf and lemon peel before serving.

· SERVING SUGGESTION ·

Sliced potatoes (1lb) can be added to the vegetable mixture after 30 minutes. Alternatively, chunks of whole wheat bread would go well with this traditional peasant dish.

INGREDIENTS

4 veal steaks (about 6oz each), trimmed of all fat

2 tablespoons olive oil

3½ cups chopped onions

3 large tomatoes, skinned and chopped or 1 small can Italian tomatoes

1lb eggplants, peeled and sliced

1lb zucchini, cut into 1-in slices

3 red bell peppers, seeded and cut into strips

1 garlic clove, crushed

½ teaspoon dried sage

1 strip lemon peel

1 bay leaf

4 servings

Picture: page 62

Calories	336	★ ★
Fat	13g	★ ★
Saturated Fat	3.0g	★ ★
Cholesterol	146mg	★ ★
Sodium	202mg	★ ★
Fiber	7.3g	★ ★

LAMB COUSCOUS WITH SEVEN VEGETABLES

for recipe see page 69

RAW LAMB PATE WITH CRACKED WHEAT

for recipe see page 81

LAMB

Lamb is rather a fatty meat, but the fat can be cut away before cooking, except the fat lies under the skin of the meat, which should always be left intact until after cooking. Lamb is very easily digestible and is therefore preferable to beef for feeding young children and others with sensitive digestions. Raw lamb should be light to dark pink, firm and fine textured. The bones should be reddened and porous. When buying standing rib or rack of lamb, ask the butcher to loosen the backbone from the ribs, to make carving easier. Allow 6oz per person for lamb off the bone and 9oz per person for lamb with the bone, such as chops and shanks.

Lamb can be left pink in the center or cooked very thoroughly, as preferred. The meat tastes very good when charred, so it is extremely tasty when barbecued, and the flavor and tenderness are enhanced by marinating before cooking.

LAMB COUSCOUS WITH SEVEN VEGETABLES

This traditional couscous dish has all the ingredients of
a balanced meal, and a little meat will serve a lot of people.
Traditionally, shoulder of mutton is used, but this meat
can be fatty when steamed in this way, so it is better to use
a leaner cut, such as leg. Couscous is prefluffed semolina. It comes
from North Africa, and can be bought in Middle Eastern grocery
stores. However, if you can't get it, use barley in the same way.
Ideally, a special pot called a *couscousière* should be used, but a
vegetable steamer set over a large casserole is a good substitute,
provided the two halves fit well together. The hot pepper sauce
called *harissa* is fairly easy to find in specialty shops, but
otherwise, there are plenty of hot pepper sauces available
commercially that will give a similar flavor.
You will need a very large serving platter, such as
a round metal tray.

· METHOD ·

Put the meat, two of the onions and the cabbage into the bottom half of
the couscous pot. Season with the saffron, salt and pepper. Add 2½
quarts water. Bring to the boil. Pour the couscous into the top half of the
pot, and cover. Cook on high heat until the steam begins to escape, then
simmer for another 30 minutes.

Remove the top half of the *couscousière* or the vegetable steamer and
cover the bottom part. Leave the vegetables and meat to simmer for at
least 1 hour. Pour the couscous into a large bowl and stir and fluff the
grains with a fork. Sprinkle them with cold water and fluff them again.
Continue sprinkling and fluffing until the grains are swollen and saturated.
Leave them to stand for at least 30 minutes. Put all the remaining
ingredients except the pumpkin into the bottom half of the pot, and
simmer for 30 minutes. Parboil the pumpkin for 15 minutes separately
and add it, then replace the couscous in the top half of the pot and simmer
for another 15 minutes.

Before serving, pile the couscous into a bowl and pour over the
cooking liquid from the meat and vegetables in spoonfuls, stirring and
fluffing the grains to help them absorb the liquid.

Arrange the vegetables around the couscous on the tray and place the
meat attractively over the top.

· SERVING SUGGESTION ·

Mix some hot sauce with a few tablespoons of cooking liquid and serve it
separately. The rest of the cooking liquid should be served unseasoned as
a gravy. Traditionally, large amounts of fat and butter are used to cook
couscous, and it is served with a lump of butter on the top, but none of
this is necessary. However, if you prefer a "buttery" finish, use low-fat
yogurt or low-fat margarine.

INGREDIENTS

2lb boned leg of lamb, trimmed of all fat and cut into large pieces
3 large onions, sliced
1 white cabbage, cut into large chunks
½ teaspoon saffron strands
½ teaspoon salt
½ teaspoon black pepper
2lb couscous
4 large tomatoes, peeled and quartered, or 1 large can tomatoes
8oz eggplant, peeled and cut into chunks
1lb carrots, sliced
1lb small turnips, trimmed and quartered
2 red bell peppers, seeded and chopped
6 tablespoons chopped fresh coriander
1 tablespoon ground coriander
1 tablespoon ground cumin
2 tablespoons harissa or other hot pepper sauce
1lb pumpkin, cut into large chunks

10 servings
Picture: page 66

Calories	462	★★
Fat	10g	★★
Saturated Fat	4.4g	★★
Cholesterol	79mg	★★
Sodium	268mg	★★
Fiber	8.1g	★★★

CROWN ROAST OF LAMB WITH KUMQUATS

for recipe see page 73

INGREDIENTS

2 limes or 1 pomegranate

12oz ground lamb

1¼ cups low-fat yogurt

1 teaspoon ground anise

1 tablespoon oil

4 cloves, bruised

1 cinnamom stick

3 cardamom pods, bruised

2 bay leaves

3oz shelled pine nuts (pignolas)

⅓ cup blanched almonds

2½ cups long-grain brown rice,
    soaked in water for 1 hour
    and drained

½ teaspoon salt or Creole Spice
    (page 27)

½ cup golden raisins

2 cups prunes, pitted and chopped

2 cups dried apricots, chopped

4 large green apples

8 servings

Picture: page 75

Calories	559	★
Fat	18g	★
Saturated Fat	2.7g	★ ★
Cholesterol	37mg	★ ★ ★
Sodium	654mg	★
Fiber	12.0g	★ ★ ★

FESTIVAL PILAF WITH GROUND LAMB

Pilaf is basically a rice dish in which the rice is first lightly fried, then steamed. A variety of delicious things may be added, like nuts and raisins. It is eaten in one form or another throughout Asia, from the Middle East to India.
This particular version is from Turkestan in central Asia. It is a complete meal in itself; the tiny amount of ground lamb it contains makes it just meaty enough to be wholly satisfying.
It is always served on special occasions, and the apples holding candles make it into a feast. The lights are turned out and the dish is brought in with the candles lit.

· METHOD ·

Squeeze the limes into a bowl, or put the pomegranate seeds into the bowl, being careful to discard all the connective tissue. Toss the lamb in the lime juice or seeds. Stir the yogurt and anise into the lamb and leave to marinate while you prepare the rest of the dish.

Heat the oil in a large nonstick skillet. Add the cloves, cinnamon, cardamom, bay leaves, pine nuts and almonds and stir-fry until the ingredients give off an aroma. Add the rice and continue stir-frying until the grains are transparent, about 8 minutes.

Transfer the mixture to a deep saucepan and add 1 quart boiling water. Put a double layer of cheesecloth over the top of the pan, then add the pan lid, making sure it is a tight fit. Steam the rice over very low heat for 15 minutes.

While the rice is cooking, season the lamb with the salt or Creole Spice. With wetted hands, roll it into tiny thumbnail-sized balls. Sauté these lamb balls in the nonstick skillet until they are no longer red on the outside.

Add the meatballs, golden raisins, prunes and apricots to the rice mixture. Stir well, cover with another paper towel and the lid, and cook for another 15 minutes.

To serve, hollow out the centers of the apples, so that the fruit is cup-shaped. Place a candle in each apple, and arrange the apples around a serving platter. Pile the rice into the center of the platter and light the candles just before serving.

· SERVING SUGGESTION ·

Like all rice dishes, pilaf can also be easily molded into a shape, such as a ring. Brush a ring mold with oil and press the cooked pilaf into it. Unmold just before serving, and fill the center of the ring with green leaves, such as watercress or lamb's lettuce.

CROWN ROAST OF LAMB WITH KUMQUATS

This elegant dish is ideal for a dinner party. By drip-roasting it over a trivet, as much fat as possible is drained away. The stuffing should be cooked separately, so that it does not absorb the fat. The lamb in this recipe is well-done, as this will also reduce the fat content; if you like it pink in the middle, reduce the roasting time by 20 minutes.
The citrus fruits provide a zesty touch, and kumquats are the only citrus fruits that can be eaten whole, skin and all.

· METHOD ·

Preheat the oven to 375°F. Curve the two racks of lamb into a ring, bone ends upward. Tie the crown roast with string to secure. Sprinkle the meat with the Creole Spice and tarragon. Arrange the roast on a trivet or a rack over a roasting pan. Crumple a large piece of foil into a ball and place it loosely in the center of the crown roast.

Roast the meat in the oven for 30 minutes. Meanwhile, reserve 14 of the kumquats. Put the rest into a food processor or blender with the parsley, green onions, mushrooms, fresh herbs, breadcrumbs and the egg. Process until the mixture is coarsely ground. Transfer it to a bowl and knead it into a ball. If it does not stick together, add a tablespoon of water.

Remove the meat from the oven. Place the meat on a dish. Discard the ball of foil and replace it with a sheet of foil large enough to fit inside the roast cavity and extend up the sides. This is so that the stuffing mixture can rest inside the cavity on the foil while cooking, and not fall through the holes in the rack. Pierce the reserved 14 kumquats at one end with a sharp knife and stick each one on a bone end of the meat. Return the roast to the rack. Put the stuffing in the center of the meat. Return the meat to the oven and roast it for a further 15 minutes.

When the roast is ready, slide it off the rack and onto a warmed serving platter. Garnish with the orange wedges and mint sprigs. Take out the stuffing, discard the foil, and pile the stuffing back into the center of the roast. To serve, carve into cutlets, and allow two cutlets and 2 tablespoons of stuffing per person.

· SERVING SUGGESTION ·

This dish is excellent with potatoes, brushed lightly with oil and roasted around the meat. A raw leafy vegetable – such as raw spinach and mushrooms, or a Bibb lettuce and watercress salad – would provide an excellent accompaniment.

INGREDIENTS

2 racks of lamb of 7 small cutlets each, trimmed of all fat, and bone ends trimmed of meat

1 teaspoon Creole Spice (page 27)

½ teaspoon dried tarragon

6oz kumquats

2 tablespoons chopped parsley

1½ cups green onions, chopped

1½ cups mushrooms, chopped

2 tablespoons fresh herbs (mint, chervil, rosemary, marjoram, etc.), chopped

½ cup whole wheat breadcrumbs

1 egg

Orange wedges and mint sprigs to garnish

6 servings

Picture: page 70

Calories	396	★ ★
Fat	20g	★
Saturated Fat	8.8g	★
Cholesterol	408mg	★
Sodium	224mg	★ ★
Fiber	4.9g	★ ★

WEDDING SOUP
for recipe see page 77

FESTIVAL PILAF WITH GROUND LAMB

for recipe see page 72

INGREDIENTS

*1lb lean lamb, cut from the leg,
   cubed*

4 tablespoons chopped parsley

2 large garlic cloves, chopped

6 green onions, chopped

1 teaspoon dried oregano

¼ teaspoon cinnamon

*½ teaspoon salt or salt substitute
   (page 26)*

½ teaspoon freshly ground black pepper

*2oz pine nuts (pignolas)
   or shelled pistachio
   nuts, skinned and split*

¼ cup cornmeal or semolina

*¼ recipe whole wheat bread dough
   (page 28) or white bread dough
   (page 29)*

8 servings

Calories	*459*	★ ★
Fat	*11g*	★ ★
Saturated Fat	*2.7g*	★ ★
Cholesterol	*44mg*	★ ★ ★
Sodium	*228mg*	★ ★
Fiber	*9.7g*	★ ★ ★

INGREDIENTS

*1 leg of lamb (about 6½lb),
surplus fat removed*

6 garlic cloves, sliced in half lengthwise

12 sprigs of rosemary

2 cups strong black coffee

½ teaspoon salt

½ teaspoon black pepper

8 servings

Picture: page 78

LAMB PIZZA WITH PINE NUTS

This is a very popular snack in the Middle East and is gaining ground in the United States, where it has been introduced by the Armenians. It makes a lovely dinner party starter, and is also ideal for children's finger food. When not for special occasions, other kinds of nuts can be substituted for the pine nuts and pistachios.

· METHOD ·

Put the lamb into a food processor and grind it finely, adding the parsley, garlic and green onions while the machine is running. Transfer the mixture to a nonstick skillet and sauté, stirring constantly to break up the pieces. Sprinkle with the oregano, cinnamon, salt, and pepper while cooking and when the mixture is no longer red, add the pine nuts, if using. Cook them until they give off their aroma, about 5 minutes. If using pistachio nuts, cook the meat for 5 minutes longer and add them at the end.

Dust 2 cookie sheets with the cornmeal or semolina. Divide the dough into 8 pieces. Roll out each piece into a circle about 6in in diameter. Arrange them on the cookie sheets. Pinch the outer edges into a rim with your fingertips and sprinkle the center of the dough circles evenly with the mixture. Bake them – in batches if necessary – in a preheated 450°F oven for 15 minutes. The meat should be sizzling. Serve hot.

· SERVING SUGGESTION ·

A particularly good salad to serve with these pizzas is one consisting of equal quantities of sliced or shredded leaf vegetables – such as bok choy, Chinese cabbage or red cabbage – and a little grated carrot. It can be bound with low-fat yogurt or Special Low-Fat Mayonnaise (page 26).

LEG OF LAMB WITH COFFEE

**This is a way of roasting lamb that lets the fat drain off, but also gives you nice brown meat juices. Normally, lamb juices look rather insipid. The coffee does not flavor the meat but improves the color of the roast itself, as well as the juices.
When removing the fat, do not touch the skin side of the meat.**

· METHOD ·

With a sharp knife, make 12 slits all over the lamb. Insert a sprig of rosemary and a piece of garlic into each slit. Place the lamb on a trivet in a roasting pan. Pour half the coffee over it. Season with salt and pepper. Place the lamb in a preheated 450°F oven and roast it for 40 minutes. Reduce the heat to 325°F and continue roasting for 2 hours.

Remove the lamb from the oven and let it rest for 10 minutes. Strain off the cooking juices and pour them into a bowl set over ice cubes. As the fat congeals on the surface, skim it off with a metal spatula and paper towels. Pour it into a saucepan and add the rest of the coffee. Bring to just below the boil. Pour into a gravy boat and serve it separately. When carving the lamb, be sure to remove all surplus fat from the slices.

· SERVING SUGGESTION ·

Potatoes roasted in the pan with the lamb are a long-standing favorite, although high in fat. Boiled potatoes and steamed brussels sprouts or cabbage wedges and carrots also go well with lamb. For a more unusual alternative you could try serving a whole grain, such as barley, soaked in water to cover for 1 hour, then baked with mushrooms; put it in the oven when you reduce the oven heat for the last hour of cooking.

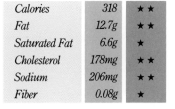

Calories	318	★ ★
Fat	12.7g	★ ★
Saturated Fat	6.6g	★
Cholesterol	178mg	★ ★
Sodium	206mg	★ ★
Fiber	0.08g	★

WEDDING SOUP

**This recipe is adapted from a special-occasion soup popular in western Asia.
The soup often contains marrow bones and is thickened with egg yolk, but both these ingredients are high in fat and have been replaced here. The soup is just as delicious without them.**

· METHOD ·

Marinate the meat and bones in the wine and vinegar, thyme, oregano and coriander for 8 hours or overnight. Transfer the meat, bones and marinade to a large casserole and add the carrot, onion, turnips and 3 quarts water. Bring to the boil. Reduce the heat and simmer for 1 hour, uncovered, skimming off any scum that rises to the surface.

Remove the meat and bones from the casserole and discard the bones. Cut the meat into ½-in wide strips. Return the meat to the casserole. Cover the pot and cook on very low heat for 2 hours. Remove ¼ cup broth from the casserole and place in a small bowl. Allow it to cool for 5 minutes, then stir the arrowroot or potato flour into it until smooth. Pour this mixture back into the casserole, stirring well. Cook, uncovered, stirring occasionally, until the soup is just below boiling point. Serve immediately.

· SERVING SUGGESTION ·

This soup is so hearty it makes a meal in itself, especially if a few boiled potatoes or cooked whole grains – such as barley or buckwheat (kasha) – are added 30 minutes before serving. Or it can simply be eaten with chunks of whole wheat bread.

INGREDIENTS

1lb lean, boneless lamb (shoulder
    or boned chops)
2lb lamb bones
2 cups dry red wine
1 cup red wine vinegar
2 teaspoons dried thyme
2 teaspoons dried oregano
½ teaspoon ground coriander
1 large carrot, split lengthwise
1 large onion, chopped
2 turnips, peeled and diced
1 tablespoon arrowroot or potato flour
1 teaspoon black pepper

10 servings

Picture: page 74

Calories	120	★ ★ ★
Fat	4g	★ ★ ★
Saturated Fat	1.9g	★ ★ ★
Cholesterol	36mg	★ ★ ★
Sodium	64mg	★ ★ ★
Fiber	0.8g	★

LEG OF LAMB WITH COFFEE
for recipe see page 76

INGREDIENTS

4 tablespoons chopped fresh mint

*¼ recipe whole wheat bread dough
(page 28) or white bread dough
(page 29)*

2lb lean lamb, cubed

½ teaspoon salt

½ teaspoon white pepper

1½ cups mushrooms, chopped

1½ cups peas

*6 medium carrots, sliced into
matchstick strips*

8 small white onions

1 egg white

6 servings

Calories	*533*	★
Fat	*16g*	★
Saturated Fat	*6.5g*	★
Cholesterol	*132mg*	★ ★
Sodium	*571mg*	★
Fiber	*12.0g*	★ ★ ★

QUICK MINTED LAMB PIE

**Pastry of all kinds needs quite a lot of fat to make the mixture stick together. Bread dough, on the other hand, needs none at all. So if you are concerned about fat content in your food, you should use fat-free yeast doughs to make pies.
Yeast doughs were used in this way in centuries past but have fallen out of fashion. If you can't be bothered to make the dough specially, frozen bread or pizza dough can be bought, but read the ingredients list to make sure it is a fat-free yeast dough. The carrots, onions and peas can all be bought frozen. The lamb mixture can be made well in advance, then chilled or frozen, and reheated and added to the pie when required.**

· METHOD ·

Work 3 tablespoons of the mint into the bread dough. Roll out the dough with a rolling pin into an 8-in circle to fit a lightly oiled nonstick pie pan. Cover it lightly with a cloth while you make the lamb mixture.

Season the lamb with salt and pepper. Put it into a nonstick skillet with a lid. Cook, stirring constantly, until it is no longer red on the outside. Add ⅔ cup water and the mushrooms and cover the pan. Simmer for 30 minutes.

Heat a pot of boiling water and quickly parboil the peas, carrots and onions – or better still, mix them in a dish, cover with plastic wrap, and microwave on high for 3 minutes.

Brush the piecrust with the egg white, and bake it in a preheated 425°F oven for 15 minutes. Pour the lamb mixture and cooking juices into the center of the pie dish, and arrange the vegetables around it. Serve immediately.

· SERVING SUGGESTION ·

If you like a thicker gravy, some of the juices can be strained off and thickened with a tablespoon of arrowroot, potato starch or cornstarch. The pie tastes best when it is served piping hot. The vegetables can be varied to include brussels sprouts, green beans, etc. A mashed root vegetable, such as parsnips or turnips, also goes well with the pie.

RAW LAMB PATE WITH CRACKED WHEAT

Raw food retains all its natural vitamins and minerals which are often destroyed during cooking. Lamb, being a relatively easily digested red meat, is surprisingly good eaten raw.
Cracked wheat is high in vegetable protein and fiber and makes an ideal carbohydrate combination with lamb. Naturally, when preparing raw food you must ensure that all the utensils and implements you use are scrupulously clean and that the ingredients are never left uncovered or in a warm place.
When this Lebanese dish is made in the traditional way, the meat is pounded in a mortar with olive oil to make it into a smooth paste. However, when a food processor is used to do the same job, it will make the mixture smooth without the need for oil. You can sprinkle the finished dish with extra virgin olive oil for decoration and authenticity, if you are not worried about increasing the fat content.

· METHOD ·

Soak the cracked wheat or bulgur in plenty of water for at least 1 hour. Squeeze any excess water from the wheat. Put the lamb cubes into a food processor and grind them with the metal blade. Then add the onion and mint leaves and grind again until you have a smooth paste. Add the salt, pepper and spices and process a third time. Transfer the paste to a bowl.

Briefly dry-fry the pine nuts in a skillet, removing them as soon as they start to give off an aroma. Stir half the pine nuts into the meat paste with the cracked wheat.

Arrange the meat neatly in an attractive serving dish. With the tip of a sharp knife, trace crisscrossing lines over the paste, and arrange the reserved pine nuts on top in a decorative pattern. Add parsley sprigs for decoration. Chill until required.

· SERVING SUGGESTION ·

Serve this as you would a pâté – as a starter or as part of an hors d'oeuvres assortment. Eat it by scooping it up with whole wheat pita or crackers.

INGREDIENTS

1lb cracked wheat or bulgur

1lb cubed lamb, cut from leg, trimmed of fat

1 onion, coarsely chopped

6 fresh mint leaves

½ teaspoon salt

½ teaspoon black pepper

⅛ teaspoon ground coriander

⅛ teaspoon ground cumin

2oz shelled pine nuts (pignolas)

Parsley sprigs to garnish

8 servings
Picture: page 67

Calories	337	★ ★
Fat	10g	★ ★
Saturated Fat	2.5g	★ ★
Cholesterol	44mg	★ ★ ★
Sodium	151mg	★ ★ ★
Fiber	0.9g	★

SHOULDER OF LAMB WITH APRICOTS

for recipe see page 84

INGREDIENTS

2 shank or knuckle ends of lamb
(4½lb)

8 sprigs of thyme

8 bay leaves

Juice and rind of 1 lemon

8 servings

Calories	259	★ ★
Fat	14g	★ ★
Saturated Fat	6.6g	★
Cholesterol	126mg	★ ★
Sodium	141mg	★ ★ ★
Fiber	0.0g	★

INGREDIENTS

1 shoulder of lamb
(about 2½lb)

1½ cups long-grain brown rice

1¼ cups dried apricots, chopped

½ cup golden raisins

½ cup chopped almonds

½ teaspoon ground cinnamon

½ teaspoon ground ginger

Whole fresh or dried apricots and
chopped fresh parsley for garnish

6 servings

KLEFTIKO
(Rustler's Lamb)

**Old-time sheep rustlers had to cook their
meat carefully to prevent the cooking smells from escaping and
giving the game away. This meant wrapping the meat in
leaves and burying it in a pit lined with hot coals.
In this modern-day version, the meat is wrapped
in foil and cooked in the oven, sealing in all the goodness
without added fat.**

· METHOD ·

Heat the oven to 350°F. Lay the shank or knuckle ends on 2 large sheets of foil. Put 4 sprigs of thyme and 4 bay leaves in with each lamb piece. Divide strips of lemon rind between the pieces of lamb and wrap them around the meat. Sprinkle each shank with the lemon juice and wrap the meat into neat packages, allowing a little room inside for the build up of steam.

Put the lamb packages into the center of the oven and cook for 2 hours. Take out one of the packages, open it carefully to avoid scalding yourself with the steam, and stick a fork or skewer into the center. If the juices run clear, the meat is done. Discard the thyme, bay leaves and lemon peel. Carve the meat and arrange it on a serving platter.

· SERVING SUGGESTION ·

This is traditionally served with oven-roasted potatoes. For nongreasy "roast" potatoes, parboil them in their skins until soft, then peel and slice them and arrange them in a shallow dish. Brush them with a few drops of oil and brown them under a hot broiler. Before serving, sprinkle the potatoes and lamb with fresh parsley. Serve with a salad of small leaves, such as Bibb lettuce, curly endive, fresh coriander and parsley.

SHOULDER OF LAMB WITH APRICOTS

**Shoulder of lamb is sometimes considered a fatty meat.
However, removing the bones gives you access to the pockets of
fat, which can be cut out at the same time.**

· METHOD ·

First bone the shoulder of lamb. Lay the meat skin side down and cut through the flesh covering the blade bone, starting where the bone joins the middle bone and cutting right to the outer edge.

Cut the meat away from the surface of the bone on either side of the first cut, folding it back to right and left, so the bone is exposed. Hold the flesh back while you cut around the bone to free it. Cut the flesh away from under it and sever the tendons. Bend the shoulder to loosen the ball-and-socket joint and pull out the blade bone. Cut away the tendons from the middle bone, cut around it and pull it out. Finally, draw back the flesh to expose the knuckle bone. Pull this out too. This will leave a deep pocket in the flesh. Cut away all excess fat, but leave the skin intact.

Bring a large saucepan of water to the boil and add the rice. Cover tightly and simmer it for 15 minutes, or until it is *al dente*. Drain and rinse, and transfer the rice to a bowl; stir in all the remaining ingredients, except the garnish. Preheat the oven to 375°F.

Stuff as much of the mixture as you can into the lamb cavity. Roll up the shoulder tightly and truss with string. Roast it for 30 minutes, then calculate another 30 minutes per 1lb stuffed weight.

To serve, arrange the meat on a serving platter, with the rest of the stuffing mixture around it. Add the whole apricots and parsley for garnish.

· SERVING SUGGESTION ·

Coleslaw, bound with low-fat dressing or Special Low-Fat Mayonnaise (see page 26), would go well with this dish.

Picture: page 82

Calories	362	★ ★
Fat	13g	★ ★
Saturated Fat	4.4g	★ ★
Cholesterol	77mg	★ ★
Sodium	403mg	★
Fiber	4.6g	★ ★

LAMB PACKETS IN MINTED YOGURT SAUCE

**This subtly flavored Middle Eastern dish looks like a sort of dumpling, but it is eaten as a main course.
The clever way in which the yogurt is made into a thick white sauce can be adapted to other dishes. The yogurt sauce would make a delicious, low-fat accompaniment for vegetables, for instance.**

· METHOD ·

Grind the lamb, onion and spices in a food processor. Mix the whole wheat flour and salt with ¾ cup water. Knead until smooth, then let rest for 10 minutes.

Roll out the dough, and use a glass or cookie cutter to cut it into 2-in rounds. Place a teaspoon of the meat mixture onto each of the rounds and draw up the dough around the meat. Moisten the edges to seal firmly.

To make the sauce, beat the yogurt with the flour and egg white, using a whisk or eggbeater. Pour the liquid into a pot and bring to the boil. Add the garlic and mint, and then the lamb packets. Reduce the heat and simmer the packets, with the pan partially covered, for 20 minutes. Serve hot.

· SERVING SUGGESTION ·

No extra starch is required, but these meatballs would go well with squash, pumpkin or broccoli, and perhaps very small helpings of brown rice. A lettuce salad would complete the meal.

INGREDIENTS

1lb lean lamb, coarsely ground
1 onion, chopped
½ teaspoon ground coriander
½ teaspoon ground allspice
3 cups whole wheat flour
¼ teaspoon salt

FOR THE YOGURT SAUCE

6 cups low-fat plain yogurt
1 tablespoon unbleached white flour
1 egg white
2 garlic cloves
*2 tablespoons chopped fresh mint or 2
   teaspoons dried mint*

6 servings

Calories	455	★ ★
Fat	10g	★ ★
Saturated Fat	4.3g	★ ★
Cholesterol	77mg	★ ★
Sodium	334mg	★ ★
Fiber	6.0g	★ ★

STIR-FRIED LEAN PORK WITH CHINESE VEGETABLES
for recipe see page 92

BRAISED PORK WITH BLACK BEANS AND ORANGES

for recipe see page 89

PORK AND HAM

When buying pork, always get the leanest cut you can.
The flesh should be greyish pink and the fat solid and white.
As soon as you get it home, unwrap the meat and rewrap it
loosely in paper before refrigerating it in the
coldest part of the refrigerator.
Always cook pork very thoroughly; it is dangerous to
undercook it. If you are cooking a pork roast, use a meat
thermometer to check the temperature. It should
register at least 170°F.
Ham and pork products are cooked but can deteriorate
quickly. For reasons of hygiene, do not use the same knife to
slice ham as you use for raw meat, without first washing it.
Ham should be treated like pork, as far as quick refrigeration
and eating are concerned.

BRAISED PORK AND BLACK BEANS WITH ORANGES

This recipe is loosely based on a South American one but uses fresh instead of salt pork. Salt pork has always featured prominently in tropical cooking because salting was the only way to preserve the meat. Prolonged soaking will remove the excess salt from the meat but also reduce the flavor, so nowadays, when there is no need to use salt meat, choose fresh instead.
Tenderloin is the best choice, since this is one of the most tender and leanest pork cuts. A cheaper alternative is loin.
If this is used, be sure to trim all the surrounding fat away from the meat. The combination of pork and beans, which contain protein but are also very starchy, is always a popular one. Black beans add a new twist, and oranges add piquancy.

· METHOD ·

Wash the beans and soak them for 2 hours or overnight in water to cover. Put the pork cubes into a large pot with the garlic, soy sauce, pepper and vinegar, and marinate for 2 hours. Drain the beans and add fresh water to cover. Put them into a large casserole and bring them to the boil. Cover the casserole and simmer the beans for 1½ hours.

Cover the pot containing the pork and simmer it on low heat for 30 minutes, or until it is tender. Remove the garlic and reserve it. Heat the oil in a skillet and add the garlic and the onion to the skillet. Fry the onion until it is transparent, then add the pork pieces and fry them. When the beans have been cooking for 1½ hours, drain off any oil from the pork and onions and transfer the contents of the skillet to the casserole in which the beans are cooking. Add the broth and simmer on a low heat for another 30 minutes.

Peel and chop the Spanish onion. Slice the oranges crosswise and arrange them in 4 side dishes. Sprinkle them with the onion and serve as a side dish to the pork stew.

· SERVING SUGGESTION ·

It is important to eat this stew with a starchy vegetable so that the bean protein can be digested. Mashed, boiled or puréed potatoes, or even baked potatoes, would be appropriate.

INGREDIENTS

8oz black beans

1lb pork tenderloin or loin, trimmed of all fat, cut into cubes

4 garlic cloves, crushed

2 tablespoons soy sauce

½ teaspoon black pepper

6 tablespoons white wine or cider vinegar

1 white onion, finely chopped

1 tablespoon oil

1 cup chicken broth (page 24)

1 Spanish onion

2 large navel oranges

4 servings

Picture: page 87

Calories	422	★ ★
Fat	13g	★ ★
Saturated Fat	3.7g	★ ★
Cholesterol	89mg	★ ★
Sodium	177mg	★ ★ ★
Fiber	18.0g	★ ★ ★

PORK BROCHETTES WITH SATAY SAUCE

for recipe see page 96

MEXICAN PORK STEAKS
for recipe see page 92

INGREDIENTS

*8oz lean pork, trimmed of all
    visible fat*

1 tablespoon soy sauce

1 tablespoon rice wine or dry sherry

1 tablespoon cornstarch

*¼ cup wood ear mushrooms or other
    dried mushrooms, soaked for
    20 minutes*

1 cup cauliflower flowerets

1 tablespoon oil

*2 green onions, sliced in half
    lengthwise*

1 cup bamboo shoots, sliced

1 cup bean sprouts

¾ cup water chestnuts, sliced

4 servings

Picture: page 86

Calories	180	★ ★ ★
Fat	8g	★ ★
Saturated Fat	2.2g	★ ★
Cholesterol	49mg	★ ★ ★
Sodium	59mg	★ ★ ★
Fiber	1.3g	★

INGREDIENTS

6 large oranges

1 tablespoon orange bitters

*8 boneless pork loin steaks (6oz
    each), trimmed of all fat*

1 teaspoon chili powder

1 teaspoon Dijon-style mustard

3 drops Tabasco sauce

¼ teaspoon ground cinnamon

8 servings

Picture: page 91

STIR-FRIED LEAN PORK WITH CHINESE VEGETABLES

**The Chinese, whose main meat is pork, very often stir-fry it.
This excellent cooking method is very quick and uses little fat. The
meat itself should be as lean as possible.**

· METHOD ·

Slice the pork into matchstick strips. Combine the soy sauce, rice wine or
dry sherry and half the cornstarch in a bowl. Use the rest of the cornstarch
to make a paste with 3 tablespoons water; reserve it.

Drain the wood ears and discard any hard parts. Chop them. Cut the
cauliflower flowerets into small pieces.

Heat a wok or large skillet and add the oil. Add the pork in its marinade
and stir-fry it for 2 minutes. Remove it with a slotted spoon.

Add the onions and cauliflower flowerets to the wok or skillet and stir-
fry for 2 minutes. Stir in the bamboo shoots and bean sprouts, chopped
dried mushrooms and water chestnuts. Stir-fry for 1 minute, return the
pork to the skillet and stir-fry; cook until liquid thickens. Serve very hot.

· SERVING SUGGESTION ·

Small pickled vegetables, such as baby gherkins, go well with this dish.
Serve it with short-grain steamed brown rice (always use short-grain rice
when eating with chopsticks), or Japanese *soba* (buckwheat noodles).

MEXICAN PORK STEAKS

**Other lean pork cuts can be used, such as chops from which
all fat and the bones have been removed. The orange juice helps
to tenderize the meat fibers.
If bitter oranges are available, use them
and omit the orange bitters.**

· METHOD ·

Grate the rind and squeeze the juice from 4 of the oranges. Add the bitters
to the liquid. Slice the other 2 oranges crosswise into 8 slices, discarding
both ends, as they contain little orange pulp.

Sprinkle the meat with the chili powder. Put the meat into a heated
nonstick skillet with a lid and cook until brown on both sides. Pour the
orange juice over the meat and sprinkle with the mustards, Tabasco and
cinnamon. Cover and simmer for 20 minutes. Add half the orange slices
and cook for another 5 minutes.

To serve, arrange a pork steak on each dish. Place an orange slice or

Calories	300	★ ★
Fat	13g	★ ★
Saturated Fat	4.7g	★ ★
Cholesterol	138mg	★ ★
Sodium	140mg	★ ★ ★
Fiber	2.2g	★

each piece of pork. Pour the sauce over them.

· SERVING SUGGESTION ·

Serve with fresh, hot corn on wheat tortillas. A *salsa*, made by chopping raw tomatoes with raw onion and sprinkling liberally with chopped fresh coriander, is the best vegetable accompaniment.

TROPICAL STUFFED ROAST PORK

Pork loin is the best and leanest cut to use for this recipe. The stuffing may also be used for suckling pig, which is leaner than mature pork. Papaya is a natural meat tenderizer; much of the enzyme which tenderizes the meat is in the seeds, which are peppery and add flavor to the roast. Plantains are black-skinned when ripe and look rather unappetizing. If they are unavailable you can substitute the same weight (about 8oz) of yams or breadfruit, but slice them and boil them first for 20 minutes.

· METHOD ·

Put the pork into a dry nonstick frying pan and brown it all over on medium heat. Remove it, leaving the fat in the pan. Add the plantain slices to the pan and sauté them, then remove and chop them finely.

Tear the bread into small pieces. Combine the orange and lime juices and soak the bread in the mixture. Add the chili powder, nuts, mango and papaya. Grind the tablespoon of papaya seeds in a food processor and stir them into the mixture. It should be dry enough to pat the stuffing into a ball; if it is not, add a handful of fresh breadcrumbs. If it is too dry, add more lime juice.

Preheat the oven to 400°F. Lay the loin out on a board and cover the surface with the stuffing mixture. Roll it up and tie it securely with string. Stick skewers in each end to ensure the stuffing does not escape.

Put the loin on a trivet in a roasting pan with ½ cup water. Roast the meat, turning every 10 minutes, until evenly browned (about 30 minutes). Reduce the heat to 325°F, and continue cooking for 2 hours or until the meat tests done when pierced with a skewer.

· SERVING SUGGESTION ·

The most suitable accompaniments are the most authentic, namely baked, boiled or steamed sweet potatoes and/or steamed yam flour (*gari*). A green salad is also a good idea. Decorate the dish with a few small chili peppers.

INGREDIENTS

5lb boneless pork loin, trimmed of surplus fat

2 ripe plantains, sliced

4 slices whole wheat bread, crusts removed

Juice and grated rind of 1 orange

Juice and grated rind of 1 lime

1 teaspoon chili powder

5oz cashew nuts, finely chopped or ground

1 ripe mango, peeled and coarsely chopped

1 ripe papaya, peeled and coarsely chopped

1 tablespoon papaya seeds

10 servings

Calories	469	★ ★
Fat	22g	★
Saturated Fat	6.1g	★
Cholesterol	158mg	★ ★
Sodium	222mg	★ ★
Fiber	3.5g	★ ★

WHOLE WHEAT HAM LOAF WITH GARDEN HERBS
for recipe see page 97

INGREDIENTS

4oz dry-roasted peanuts

½ small onion

1 garlic clove

1 tablespoon soy sauce

3 tablespoons vinegar

1½ teaspoons lemongrass

2 tablespoons ground coriander

1 tablespoon brown sugar

*¼ teaspoon salt or salt substitute
(page 26)*

¼ teaspoon black pepper

¼ teaspoon cayenne pepper

1½lb lean pork cubes, trimmed of fat

2 cucumbers

4 tomatoes

8 servings

Picture: page 90

Calories	*233*	★ ★ ★
Fat	*13g*	★ ★
Saturated Fat	*3.6g*	★ ★
Cholesterol	*74mg*	★ ★ ★
Sodium	*186mg*	★ ★ ★
Fiber	*1.7g*	★

PORK BROCHETTES WITH SATAY SAUCE

**Boiling and barbecuing are both good ways to drain the
fat from meat while it cooks.
This Southeast Asian dish is very high in protein, supplied by both
the pork and the peanuts. For this reason, a little
goes a long way, and only small portions are needed,
about 2 small skewers per person.**

· METHOD ·

In a blender or food processor fitted with the steel blade, combine the peanuts, onion, garlic, soy sauce, vinegar, lemongrass, coriander, brown sugar, salt or salt substitute, black pepper and cayenne. Process or blend until as smooth as possible. Pour this satay sauce over the pork and cover it. Leave it to marinate for at least 3 hours at room temperature or, better still, overnight.

When ready to cook, peel the cucumbers, seed them and cut them into 1-in chunks. Cut each tomato into quarters. Thread the meat and vegetables alternately onto skewers. Cook them slowly over or under a hot broiler for 25 minutes, basting occasionally with the marinade. Serve any remaining marinade separately.

· SERVING SUGGESTION ·

The best starchy accompaniment to this dish is short-grain brown rice. Serve fresh fennel and cucumber salad, sprinkled with vinegar – especially Japanese rice vinegar – to make a very refreshing palate-lifter. You can make more sauce by grinding more unsalted dry-roasted peanuts in a blender with a little vinegar.

INGREDIENTS

*2lb pork spareribs, cut into
individual ribs, visible fat removed*

2 tablespoons molasses

2 tablespoons dry sherry (optional)

2 tablespoons wine or cider vinegar

2 tablespoons soy sauce

1 teaspoon ground ginger

*14oz can crushed pineapple in
unsweetened juice*

⅓ cup dry whole wheat breadcrumbs

4 servings

CARIBBEAN SPARERIBS WITH FRUITY SAUCE

**Pork spareribs must be treated with caution as they are
very fatty, but if they are barbecued or broiled the fat can
drain away. Children love them.
This sauce is based on a West Indian recipe but with a few
adjustments, such as leaving out the coconut. Coconut, although
delicious, is high in saturated fats and best eaten infrequently. You
can substitute a commercial steak sauce for the soy sauce.**

· METHOD ·

Heat a large pan of water to the boiling point. Add the ribs and simmer for 30 minutes. Drain well. Combine the molasses, sherry – if using – vinegar, soy sauce, ginger, crushed pineapple and breadcrumbs. Bring to the boil and simmer the mixture, stirring occasionally, for 5 minutes.

Arrange the ribs on cookie sheets. Brush each rib with the sauce and place under a hot broiler. Broil for 10 minutes. Turn and broil for 8 minutes. Keep brushing the sauce onto the ribs to keep them moist. Serve hot with the rest of the sauce.

· SERVING SUGGESTION ·

A starchy vegetable, such as baked sweet potatoes, would go very well with these ribs. Yams would be nicest; boil them like ordinary potatoes, but slowly to bring out the flavor. A salad in which leaf vegetables such as lettuce and cabbage are combined with fresh fruits, such as apples, oranges, bananas and plums, will round off the meal.

Calories	225	★ ★ ★
Fat	5g	★ ★ ★
Saturated Fat	1.8g	★ ★ ★
Cholesterol	49mg	★ ★ ★
Sodium	160mg	★ ★ ★
Fiber	2.4g	★

WHOLE WHEAT HAM LOAF WITH GARDEN HERBS

Meat loaf is a popular form in which to eat ground meat; this version is both low in fat and very tasty, due to the herb flavorings. Any leftovers can be shaped into patties and broiled. The use of ham, which is slightly salted, eliminates the need for added salt.

· METHOD ·

Preheat the oven to 350°F. Mix all the ingredients in a bowl until they are well blended and stick together.

Lightly oil a 2-quart nonstick loaf pan and press the ham loaf into it. Bake it for 90 minutes. Remove it from the oven and cool for 15 minutes before unmolding. Serve hot or cold.

· SERVING SUGGESTION ·

Whole wheat ham loaf goes extremely well with salad. It makes an elegant buffet dish, especially if it is carefully presliced, then returned to its original shape and decorated with strips of raw vegetables.

INGREDIENTS

1lb lean ham, ground

1lb lean pork, ground

¾ cup soft whole wheat breadcrumbs

1 egg, beaten

1 cup low-fat yogurt

1 green bell pepper, seeded and chopped

2 tablespoons chopped parsley

1 tablespoon dried sage

1 teaspoon dry mustard

Oil for greasing the loaf pan

8 servings

Picture: page 94

Calories	236	★ ★ ★
Fat	9g	★ ★
Saturated Fat	3.1g	★ ★
Cholesterol	96mg	★ ★
Sodium	899mg	★
Fiber	2.2g	★

LIGHT HAM MOUSSE

for recipe see page 100

BREAD DOUGH
for recipe see page 28/29

INGREDIENTS

*1 ready-to-eat cooked boneless ham
   (about 8lb)*

3 large grapefruit

1 cup dry white wine

6 tablespoons honey

1 tablespoon cornstarch

16 servings

Calories	358	★ ★
Fat	10g	★ ★
Saturated Fat	3.9g	★ ★
Cholesterol	144mg	★ ★
Sodium	2084mg	★
Fiber	0.2g	★

INGREDIENTS

1 package unflavored gelatin

2 celery stalks, chopped

*8 pitted olives (stuffed or not, as you
   prefer)*

¾ cup tofu, well drained

1 teaspoon paprika

1 teaspoon cayenne pepper

2 tablespoons capers

8oz cooked lean ham, ground

8 small servings

Picture: page 98

GRAPEFRUIT-GLAZED HAM

**Ham is a salty meat, so it is a good idea not to eat it too often, but
many families like to serve it on special occasions.
The bittersweet flavor of grapefruit combines very well with the
honey and the ham. This recipe is not only appetizing, but
extremely quick and easy to prepare.**

· METHOD ·

Preheat the oven to 325°F. Remove as much fat as possible from the ham, and put it on a trivet in a roasting pan, fat side up. Bake it for 1½ hours. Remove the ham from the oven and score the fat into diamonds.

    Squeeze the juice from two of the grapefruit and combine it with the wine and honey. Pour the mixture over the ham and return the meat to the oven. Bake for 40 minutes, basting every 10 minutes with meat juices. While the ham is cooking, peel and section the remaining grapefruit, discarding as much pith and skin as possible.

    Remove the ham from the oven and transfer it to a serving dish. Put the roasting pan on the stove over low heat. Combine the cornstarch with ½ cup water and add this liquid to the meat juices. Cook, stirring constantly, until the liquid thickens and boils. Pour it over the ham. Garnish the ham with the reserved grapefruit sections.

· SERVING SUGGESTION ·

For a special occasion, the ham can be additionally garnished with cooked sour cherries and a salad containing a variety of citrus fruits. As ham is traditionally a winter dish, try it with a cooked root vegetable – such as mashed rutabaga, parsnip or turnip – instead of potato, plus a cooked green vegetable, such as kale or cabbage.

LIGHT HAM MOUSSE

**This delicious starter or buffet dish is also low in fat.
Normally, mousses are made with a combination
of gelatin and cream. Tofu, which is almost tasteless, is an
excellent substitute for both eggs and cream in all kinds of dishes; it
is low in fat and high in protein and too useful to be left only
to the vegetarians!**

· METHOD ·

Sprinkle the gelatin over 6 tablespoons cold water. Leave for 5 minutes to soften. Place the mixture over a pot of boiling water, and stir until the gelatin has dissolved.

    Mix together thoroughly, either by hand or in a food processor, the celery, olives, tofu, paprika, cayenne pepper, capers and ham. Stir the gelatin into the mixture. Rinse out a mold with cold water and pile the mixture into it. Refrigerate until set, about 4 hours. Unmold before serving.

Calories	70	★ ★ ★
Fat	3g	★ ★ ★
Saturated Fat	0.8g	★ ★ ★
Cholesterol	10mg	★ ★ ★
Sodium	522mg	★
Fiber	0.9g	★

· SERVING SUGGESTION ·

Serve with whole wheat crackers, crispbread or bread, or as a starter with a few lettuce leaves. For a party, unmold the mousse and sprinkle it with more paprika; decorate with raw vegetables before serving.

CREOLE BEANS AND RICE WITH SMOKED HAM

This nutritious combination of a starch with a second-class protein is enhanced by the addition of a little smoked pork meat. The bland flavor of the rice and beans thus needs no salt. White rice will look more attractive than brown, and the dish is very nutritious in any case. If you are cooking the red kidney beans yourself, it is very important for them to boil hard for at least 10 minutes so that the toxins present in the raw bean can evaporate.

INGREDIENTS

1lb red kidney beans, soaked in water for 3 hours

1lb smoked ham hocks or steaks, all fat removed, cut into large serving pieces, bones reserved

1 large onion, chopped

1 green bell pepper, seeded and chopped

2 garlic cloves, chopped

1 bay leaf

5 drops Tabasco sauce

2 teaspoons Creole Spice (page 27)

1 cup basmati rice

4 tablespoons chopped parsley

8 servings

· METHOD ·

Drain the beans and put them into a pot with fresh water to cover. Bring to the boil, boil briskly for 10 minutes, then reduce the heat. Add the meat and bones, onion, pepper, garlic, bay leaf and seasonings. Cover and simmer for 2 hours, adding more water if the beans look dry.

Bring 3½ cups water to the boil in another pot and add the rice. Bring back to the boil, and when the rice "dances," reduce the heat and cover the pot. Cook for 12 minutes. Rinse and drain the rice and steam it in a colander over an inch of boiling water, covered with a kitchen towel to absorb the excess moisture.

Discard the bones and bay leaf. Pour the beans and meat mixture into the center of a large serving platter; arrange the rice around the edge. Sprinkle the bean mixture with the parsley.

· SERVING SUGGESTION ·

This meal has plenty of starch but would benefit from the addition of a few fresh vegetables or a green salad. Instead of the same old lettuce, why not try shredding leaf vegetables that are normally cooked, such as spring greens, or spinach and Swiss chard, and combining them with fresh herbs, such as parsley and basil? You can also add unusual fresh greens like lamb's lettuce and arugula.

Calories	361	★ ★
Fat	6g	★ ★ ★
Saturated Fat	1.8g	★ ★ ★
Cholesterol	29mg	★ ★ ★
Sodium	1082mg	★
Fiber	16.0g	★ ★

JAFFA ROAST CHICKEN
for recipe see page 105

LEMON CHICKEN SOUP WITH HERBS

for recipe see page 124

CHICKEN

Chicken has a much lower fat content than red meat, though like other fowl, the light breast meat is lower in fat than the dark leg meat. Most of the fat is concentrated around the rear and under the skin of the bird; always discard pockets of fat before cooking. Before braising or stewing a chicken, remove as much of the skin as possible and discard it; if the bird is being roasted, the weight-conscious should remove the skin from the meat after cooking. Chicken fat, however, has only half the cholesterol of butter.

When buying chicken, look for birds with plump breasts. Do not worry if the skin color is yellowed, this is sometimes caused by the diet of the bird. It is a good idea to smell the flesh; birds fed on fishmeal will smell and taste unpleasantly fishy. It is a good idea to try to find chickens that still contain the giblets; unfortunately, there is a trend among meat markets to sell the giblets elsewhere (for pet food, for instance). When preparing whole chickens for cooking, pull out all loose fat from around the vent opening. Examine the back of the tail piece to make sure the fat glands have been removed (yellow deposits on either side of the bone). Inspect the bird for stray hairs and unplucked feathers and pull them out or singe them. Remove any giblets still inside and reserve them. Wash the bird inside and out in running water and pat it dry.

Allow 6oz per serving of boneless poultry and
9oz for drumsticks and other bony portions,
or half a Rock Cornish hen per person.

JAFFA ROAST CHICKEN

This elegant treatment for a roast chicken will make it fit for a king. It is the ideal dinner party dish for those who are worried about their weight, especially if the skin is not eaten.

The skin will roast to a very dark brown; this is the effect of the basting sauce. The bird would cook very well in a microwave or combination oven, as it would take on a good color, despite the lack of conventional dry heat.

The orange used to stuff the bird must be well washed, as the cooked orange rind will be eaten, and the surface may have been treated with wax to keep it fresh in the store.

If the chicken is cooked conventionally, there will be almost no liquid left in the bottom of the pan; if it is cooked by one of the alternative methods, the liquid can be degreased by chilling and skimming off surplus fat, then served as a sauce.

INGREDIENTS

3½-lb roasting chicken, visible fat removed

2 seedless oranges, well washed

6 whole cloves

1 tablespoon honey

¼ teaspoon ginger

Juice and grated rind of 1 lemon

2 tablespoons soy sauce

½ onion, grated

6 servings

Picture: page 102

· METHOD ·

Prick the chicken skin all over and place the bird in a roasting pan on a trivet. Add 1 cup water to the pan. Stick one of the oranges with the whole cloves, evenly spaced, and stuff the whole orange inside the chicken. Squeeze the juice of the remaining orange and combine it with the honey, ginger, lemon juice and rind, soy sauce and onion. Pour this mixture over the chicken.

Leave the chicken to rest while you heat the oven to 400°F.

Roast the chicken for 30 minutes, basting at least once during that time. Reduce the heat to 350°F and continue cooking for 75 minutes, basting every 15 minutes with the liquid. Before serving, remove the orange from inside the bird and cut it into 6 pieces. Decorate each portion of chicken with an orange segment.

Calories	405	★★
Fat	28g	★
Saturated Fat	11.3g	★
Cholesterol	176mg	★★
Sodium	115mg	★★★
Fiber	0.1g	★

· SERVING SUGGESTION ·

Serve with Festival Pilaf (page 72), omitting the ground lamb from the recipe, or with brown rice and raisins (see page 108).

CHICKEN PACKETS WITH SPRING VEGETABLES
for recipe see page 120

CLAY POT CHICKEN WITH OLIVES
for recipe see page 109

INGREDIENTS

3-lb chicken, jointed

*¼ cup chick-peas (garbanzo beans),
      soaked in water to cover overnight*

4 onions, chopped

¼ teaspoon ground saffron

1¼ cups raisins

1 cup long-grain brown rice

6 servings

Calories	388	★ ★
Fat	6g	★ ★ ★
Saturated Fat	0.3g	★ ★ ★
Cholesterol	99mg	★ ★
Sodium	368mg	★ ★
Fiber	7.3g	★ ★

INGREDIENTS

*2 whole chicken breasts, skinned and
      boned (about 12oz each)*

1 cup chicken broth (page 24)

½ teaspoon ground allspice

¼ teaspoon ground ginger

2 slices crispbread (such as Ryvita)

6 tablespoons white wine vinegar

1lb fresh spinach

½ teaspoon grated nutmeg

½ cup blanched almonds

½ cup ground almonds

4 servings

BRAISED CHICKEN WITH BROWN RICE AND RAISINS

**The combination of brown rice and chick-peas (garbanzos) makes
this dish particularly rich in protein. Like all the recipes in this book
it is relatively low in fat and sodium.**

· METHOD ·

Skin the chicken and remove all visible fat. Put it into a large deep pot and
add the chick-peas, onions, and water to cover. Sprinkle with the saffron.
Bring to the boil over high heat, then reduce the heat, cover the pot and
simmer for 45 minutes. Add half the raisins and cook for another 15
minutes.

    Bring a large pot of water to the boil and add the rice. When it begins
to "dance," reduce the heat and cover the pot. Simmer for 30 minutes.
Drain the rice and rinse it briefly. Stir in the remaining raisins and transfer
to the top half of a steamer or double boiler. Reheat for 10 minutes.

    Serve the chicken on a bed of rice. Pour some of the cooking juices
over the chicken and serve the rest separately, or save it for broth.

· SERVING SUGGESTION ·

Pieces of pumpkin can also be added to the chicken during the last half
hour of cooking, to make the dish even more substantial. A green
vegetable, such as broccoli or turnip greens, would provide contrast in
both color and texture, as would garden or snow peas.

CHICKEN BREASTS WITH SPINACH AND ALMONDS

**This recipe combines spinach, which is rich in carotene, with easily
digestible chicken meat and the plant protein in the almonds.
It is a very elegant dish, and one especially suitable for people
who find red meat indigestible.
This method of thickening a sauce with ground almonds
and crumbs is at least 500 years old.**

· METHOD ·

Remove all visible fat from the chicken breasts, and slice them into
serving pieces. Put them into a saucepan with a lid and add the chicken
broth, allspice and ginger. Cover the pan and simmer the breasts for 30
minutes or until tender (they can also be microwaved for 15 minutes).
Soak the crispbread in the vinegar.

Rinse the spinach and put it into a deep pot, without adding extra water. Sprinkle it with the nutmeg and cook it on medium heat for 5 minutes, or until wilted (or microwave it for 2 minutes). Chop it coarsely, and stir the whole almonds into it. Arrange it on a serving dish.

Stir the ground almonds into the vinegar and crispbread mixture. Drain the chicken breasts and arrange them on a serving dish. Pour the ground almonds and vinegar mixture into the cooking liquid, and cook, stirring, over medium heat, until the sauce thickens. Arrange the chicken on the spinach bed and pour the sauce over the breasts.

· SERVING SUGGESTION ·

Whole wheat pasta shells would make an appetizing accompaniment, as would a whole grain such as millet or buckwheat. Allow 2oz raw weight per portion. Soba, Japanese buckwheat noodles, would also taste good.

Calories	393	★ ★
Fat	20g	★
Saturated Fat	3.0g	★ ★
Cholesterol	121mg	★ ★
Sodium	262mg	★ ★
Fiber	4.9g	★ ★

CLAY POT CHICKEN WITH OLIVES

A clay casserole with a tightly fitting lid is ideal for any kind of braising or stewing but is especially good for chicken. These pots are often shaped like a sitting hen. The clay is unglazed and is soaked in water so that it steams the food during cooking. However, any kind of stewing or steaming in a closed vessel will hold the fat, so always skin the chicken before cooking it this way. If the olives are well rinsed before cooking, they will not be too salty, and of course, they remove the need for added salt.

· METHOD ·

Soak the clay pot in cold water, then drain it. Wipe the outside with a kitchen towel. Arrange the chicken pieces and giblets, the garlic, onion and spices inside the pot. Put the pot into a cold oven and turn the heat to 350°F. Cook the chicken for 45 minutes.

Meanwhile, rinse the olives, and slice the lemons crosswise into ¼-in slices. Mix them together with the bay leaf. Add them to the pot and cook for another hour. Discard the bay leaf before serving.

· SERVING SUGGESTION ·

The rich cooking liquid should be left in the pot and can be used as a sauce. Remove the liver and use it in another dish, but pull the meat from the neck and shred it into the sauce. Save the gizzard and neck for Cajun Rice (see page 171) or another dish. Serve the chicken with brown rice or with other grains, such as barley or wheat berries, and a green salad.

INGREDIENTS

6 chicken parts (legs, breasts or wings), skinned, all visible fat removed

1 set chicken giblets (neck, liver, gizzard, heart)

2 garlic cloves

1 onion, stuck with 2 cloves

1 teaspoon Creole Spice (page 27)

½ teaspoon ground coriander

⅛ teaspoon saffron strands

1lb pitted Mission green olives

2 lemons

2 bay leaves

6 servings

Picture: page 107

Calories	209	★ ★ ★
Fat	12g	★ ★
Saturated Fat	2.6g	★ ★
Cholesterol	105mg	★ ★
Sodium	1447mg	★
Fiber	2.9g	★

CHINESE CHICKEN AND WALNUTS

for recipe see page 128

SPANISH CHICKEN WITH BROWN RICE

for recipe see page 113

INGREDIENTS

4½-lb chicken, cut into 8 serving
    pieces, skinned, all visible fat
    removed

1 medium onion, finely chopped

1¼ cups tomato juice

1 tablespoon tomato paste

3 cups chicken broth
    (page 24)

½ teaspoon black pepper

½ teaspoon ground turmeric

6 cups green beans, sliced

1 large tomato, skinned and chopped

½ teaspoon salt or salt substitute
    (page 26)

4 servings

Calories	274	★★
Fat	8g	★★★
Saturated Fat	1.8g	★★★
Cholesterol	149mg	★★
Sodium	562mg	★
Fiber	5.9g	★★

INGREDIENTS

3½-lb roasting chicken

6 tablespoons chicken broth (page 24)

1 tablespoon cornstarch

4 tablespoons lemon juice

2 teaspoons dried oregano

2 sprigs dried sage

1 teaspoon dried rosemary

2 garlic cloves, finely chopped

1 tablespoon Creole Spice (page 27)
    or chili powder

4 servings

BAKED CHICKEN RAGOUT

**This simple stew is an excellent way of making a chicken
casserole, using fresh, natural ingredients, without
the need for packaged flavorings.
If you do not want the sauce to be thick, do not bother to brown the
chicken first, but cook it in the oven for 90 minutes at 350°F in an
unglazed clay pot or a casserole with a tight-fitting lid.**

· METHOD ·

Put the chicken pieces into a nonstick skillet with a lid. Add the onion and
cook, turning frequently, until it is lightly browned all over, about 25
minutes.

    Mix the tomato juice with the tomato paste and the chicken broth.
Pour the sauce over the chicken. Bring the ragout to the boil, cover the
skillet, reduce the heat and simmer for 20 minutes. Add the rest of the
ingredients and cook for a further 30 minutes. If the liquid is not reduced
to a thick sauce, uncover the skillet and continue cooking until it thickens.

· SERVING SUGGESTION ·

Serve with chunks of whole wheat bread to mop up the sauce, or with
brown rice.

MARINATED BROILED CHICKEN

**This simple broiled chicken recipe is a wonderful way to use one of
the large oven broilers in modern ranges, or a stand-alone broiler.
Marinades often contain oil to improve the texture
of the marinade and to make sure it does not run off the food too
easily while it is broiling.
This marinade is thickened with cornstarch instead.
For those worried about their weight, the chicken skin should be
removed before serving.**

· METHOD ·

To prepare the chicken, split it in two by cutting through it from the
breastbone to the backbone. Lay the halves down, skin side up, and slice
each one through diagonally between the breast and the wing.

    In a small saucepan, warm the chicken broth. Mix the cornstarch with
2 tablespoons of water and stir it into the broth. Continue stirring until the
liquid thickens. Remove from the heat and leave to cool.

    Combine the lemon juice, oregano, sage, rosemary, garlic and Creole
Spice or chili powder. Mix this into the cooled, thickened broth. Place the
chicken pieces in a large, shallow dish and pour this liquid over them. Let
them marinate for at least 2 hours, turning at least 4 times.

Heat the broiler, with broiler pan in place underneath it, so that it also gets hot. Place the chicken pieces, skin side downward, on the broiler rack. Spoon some of the marinade mixture over them and broil the chicken for 15 minutes. Turn the pieces over and brush them with more marinade mixture. Continue broiling for 20 minutes, or until the skin is crisp, brushing at least twice with more marinade mixture.

· SERVING SUGGESTION ·

This should be a simple, hearty meal, with simple, hearty accompaniments. Serve it with a plain green salad and baked potatoes in their jackets. If the potatoes are parboiled and pierced with metal skewers, they can be put into the bottom of the oven at the same time as the chicken is placed under the broiler and should be ready to eat at the same time as the chicken.

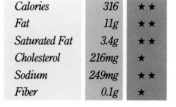

Calories	316	★★
Fat	11g	★★
Saturated Fat	3.4g	★★
Cholesterol	216mg	★
Sodium	249mg	★★
Fiber	0.1g	★

SPANISH CHICKEN AND BROWN RICE

This hearty casserole is lower in fat than the traditional Spanish version, but no less delicious for all that. Use an oil low in polyunsaturates, instead of the traditional olive oil, if you prefer. Removing the skin from the chicken before cooking not only reduces the fat content but improves the dish, as many people feel casseroled chicken skin is rubbery and unpleasant.

· METHOD ·

Put the oil into a nonstick skillet and add the chicken pieces. Cook until they are no longer red. Remove and drain them on paper towels.

Add the onion, garlic and peppers to the skillet and continue cooking until the onion is transparent, about 7 minutes. Add the tomatoes and sprinkle them with turmeric. Heat the oven to 375°F.

Transfer the chicken pieces to an ovenproof casserole. Pour the mixture from the skillet over them and add the rest of the ingredients. Cover the casserole and bake for 1 hour, checking after 30 minutes to see that there is still liquid in the bottom of the casserole. If not, add up to 6 tablespoons water or broth. Serve hot with brown rice.

· SERVING SUGGESTION ·

A salad using the same vegetables as in the stew – onions, tomatoes and peppers – but this time diced and sprinkled with lemon juice and served with Special Low-Fat Mayonnaise (page 26), is an excellent accompaniment to the dish.

INGREDIENTS

1 tablespoon oil

4-½lb roasting chicken, cut into serving pieces, skinned, all visible fat removed

1 large onion, thinly sliced

2 garlic cloves, chopped

2 red bell peppers, seeded and cut into strips

3 large tomatoes, skinned and chopped

1 teaspoon ground turmeric

Juice of 2 lemons

1 teaspoon honey

½ teaspoon cayenne pepper

6 servings

Picture: page 111

Calories	230	★★★
Fat	9g	★★
Saturated Fat	1.9g	★★★
Cholesterol	132mg	★★
Sodium	128mg	★★★
Fiber	1.8g	★

INGREDIENTS

4 tablespoons mango chutney

1 tablespoon curry powder

2½ cups low-fat yogurt

2 tablespoons cornflour

*4 chicken breasts
    (about 6oz each)*

*2 medium-sized mangoes
    (about 6oz each)*

Juice of 1 lemon

2 tablespoons slivered almonds

Extra mango slices for garnish

4 servings

Calories	751	★
Fat	43g	★
Saturated Fat	24.2g	★
Cholesterol	227mg	★
Sodium	1927mg	★
Fiber	2.4g	★

INGREDIENTS

2 chicken breasts, skinned and boned

1 cup chicken broth (page 24)

1 packet unflavored gelatin

½ teaspoon ground turmeric

*2 cups Special Low-Fat
    Mayonnaise (page 26)*

1½ cups cooked diced carrots

1½ cups cooked peas

1 cup chopped celery

4 servings

CHICKEN BREASTS WITH MANGO

This exotic dish is ideal for a dinner party, and mangoes are a good source of vitamins A and C. Some mangoes are ripe when green, but make sure they are plump, not shriveled, when they tend to become very fibrous.

· METHOD ·

Combine the chutney, curry powder and yogurt. Pour them into a pan. Mix the cornflour with ½ cup water and add to the pan. Bring to the boil, stirring continuously, until the mixture thickens. Add the chicken and cover the pan. Simmer very gently for 45 minutes.

Peel the mangoes and cut the flesh away from the stones. Add the mango flesh to the chicken mixture with the lemon juice. Simmer for 5 more minutes. Sprinkle with the slivered almonds, and garnish with more mango slices.

· SERVING SUGGESTION ·

The best accompaniment to this dish is long-grain brown rice, with a few raisins and slivered almonds added. If you are not worried by extra calories, you can also add a few slices of banana.

CHICKEN SALAD WITH SPECIAL LOW-FAT MAYONNAISE

Mayonnaise salads look so elegant in a buffet, but you do not know what is in them and, if they are made with real mayonnaise, they have a very high fat content. This salad makes an attractive buffet dish, but one that is much lower in calories. Serve it chilled, and refrigerate again immediately after serving. If you have made chicken broth that has jelled by itself (by adding the feet, for instance, as suggested) you will not need the gelatin.

· METHOD ·

Put the chicken breasts into the broth and simmer on top of the stove for 30 minutes (or microwave on high for 10 minutes). Remove the chicken breasts and leave them to cool. Soften the gelatin in 2 tablespoons of the broth for 10 minutes. Then add it to the rest of the broth and stir over low heat to dissolve. Do not let the liquid boil. Remove it from the heat, transfer it to a bowl and refrigerate until required.

Half an hour before serving, stir the turmeric into the Special Low-Fat Mayonnaise so that it looks faintly yellow. Arrange the chicken breasts on a serving dish and slice them through diagonally into serving pieces. Spread the Special Low-Fat Mayonnaise over them neatly (use a wet kitchen towel to mop up any spills). Arrange the vegetables around the chicken. If the chicken broth has jelled, heat it gently only until it liquefies again; it should only be slightly warm. Pour it over the vegetables and refrigerate until required.

· SERVING SUGGESTION ·

The coated chicken can be decorated with vegetables sliced into matchstick strips. Serve with a green salad of iceberg and oakleaf lettuce, and chicory. A cold brown rice dish would make a satisfying and nutritious accompaniment.

Calories	243	★★★
Fat	6g	★★★
Saturated Fat	2.1g	★★
Cholesterol	109mg	★★
Sodium	364mg	★★
Fiber	9.8g	★★★

CARIBBEAN CHICKEN

Limes contain an enzyme that tenderizes meat, much like pineapple and papaya. For this reason, where limes are a common fruit – particularly in the tropics – they are often mixed with meat and fish before cooking.
The classic way to make this chicken is with coconut milk, but coconut is one of the few nuts that is high in saturated fats. Consequently the coconut milk has been replaced with soy milk. The effect is the same.

· METHOD ·

Mix the soy milk with the lime juice and rind, chili pepper, bay leaf, cardamom and cloves. Put the chicken pieces into a casserole (not one made of aluminum) and add the liquid. Marinate for 1 hour at room temperature.

Heat the oven to 350°F. Cover the pot and cook the chicken for 90 minutes, or until tender. Serve it with the cooking liquid.

· SERVING SUGGESTION ·

A starchy vegetable, such as breadfruit or yucca, would be eaten with this chicken dish in the West Indies. Another excellent accompaniment is sweet potatoes, baked in their jackets while the chicken is in the oven.

INGREDIENTS

2 cups soy milk
Juice and grated rind of 2 limes
1 small whole green chili pepper
1 bay leaf
4 crushed cardamom pods
4 whole cloves
3½lb chicken pieces, skinned, all visible fat removed

6 servings

Calories	167	★★★
Fat	7g	★★★
Saturated Fat	1.9g	★★★
Cholesterol	99mg	★★
Sodium	97mg	★★★
Fiber	0.0g	★

INGREDIENTS

*1lb boneless, skinless, lean chicken
    meat*

8oz (about 8) chicken livers

¼ cup soy sauce

¼ cup dry white wine

1 teaspoon dried tarragon

1 teaspoon dry mustard

3 cups button mushrooms

*12 pearl onions (or the bulbs of large
    green onions)*

6 servings

Calories	*172*	★ ★ ★
Fat	*6g*	★ ★ ★
Saturated Fat	*1.9g*	★ ★ ★
Cholesterol	*210mg*	★
Sodium	*100mg*	★ ★ ★
Fiber	*1.4g*	★

INGREDIENTS

*12oz boneless chicken breast, skinned,
    all visible fat removed*

Juice of 2 lemons

4 large chayotes

1 cup chicken broth (page 24)

1 medium onion, chopped

½ cup chopped celery

½ cup whole wheat breadcrumbs

4 tablespoons grated Parmesan cheese

½ teaspoon cayenne pepper

½ teaspoon paprika

4 servings

MARINATED CHICKEN BROCHETTES

**Although chicken flesh is lean, it requires no added fat at all for
cooking. Naturally, the other ingredients on the skewers can be
varied to suit individual tastes.**

· METHOD ·

Have 6 skewers ready, preferably wooden ones. Cut the chicken and liver
into bite-sized pieces. Mix the soy sauce, wine, tarragon and mustard and
marinate the chicken meat and livers for at least 2 hours.

Thread the meat and liver alternately onto the skewers, with a
mushroom and an onion between each piece of meat.

Grill over a hot barbecue or under a broiler, turning and basting
frequently with the marinade ingredients, for 10 minutes.

· SERVING SUGGESTION ·

The soy sauce in the marinade would seem to suggest serving this dish
with rice, but rice is not always convenient for a barbecue. A better idea is
sweet potatoes, wrapped in foil and cooked in the coals or baked in the
oven. Parboil the sweet potatoes first to ensure quick and even roasting.

CHICKEN CHAYOTE

**One of the reasons why this pale green pear-shaped member of the
squash and cucumber family is not widely known is because it has
such a huge variety of names! It is called variously choko, xoxo,
christophene, chayote, mirliton and even vegetable pear.
It is easy to find in the Sun Belt and is used extensively in the
Creole cooking of the southeastern United States. The underrated
chayote is very valuable to those who are concerned with health,
since it has the attractive appearance of an avocado with none of
the fat and is very low in calories.**

· METHOD ·

Slice the chicken breast and stew it in a tightly covered pot with the juice
of 1 lemon on very low heat until cooked through (or microwave for 5
minutes on high in a plastic-wrap-covered dish). Leave it to cool.

Slice the chayotes in half and carefully remove the center seeds. Put
them into a large pot of boiling water and cook them until they are just
tender, about 15 minutes, testing them with a fork but being careful not to
pierce the outer skin. Scoop out most of the flesh, leaving a thin layer
inside the shell, again, being careful not to pierce the skin. Arrange the
chayotes in an ovenproof serving dish. Squeeze the other lemon and
sprinkle the juice over the chayotes. Pour the broth into the dish around
the chayotes. Put the chicken into a food processor and grind it. Add the

scooped out chayote flesh and the chopped vegetables; grind again until you have a smooth mixture. Stir in half the breadcrumbs and cheese.

Heat the oven to 350°F. Shape the mixture into 4 patties and use it to stuff the chayotes, mounding the mixture slightly. Sprinkle with the remaining breadcrumbs and cheese. Bake them for 20 minutes, or until lightly browned. Mix the cayenne and paprika and sprinkle on the chayotes before serving.

· SERVING SUGGESTION ·

This makes an elegant hot appetizer, but it can just as easily be prepared in advance and served cold with green salad. You will also most certainly have chicken mixture left over. This can be shaped into patties and heated through in a nonstick skillet, broiled or baked in a few tablespoons of broth in the oven and served with vegetables at another meal.

Calories	242	★ ★ ★
Fat	8g	★ ★ ★
Saturated Fat	3.2g	★ ★
Cholesterol	74mg	★ ★ ★
Sodium	330mg	★ ★
Fiber	3.0g	★ ★

LEMONY CHICKEN-IN-THE-POT

**Chicken-in-the-pot is usually made with a whole chicken, but it is hard to remove the fat from a whole bird.
Furthermore, the skin of braised and boiled chickens is rubbery and is not very tasty, so skinned chicken parts will actually taste better and, at the same time, be lower in fat.**

· METHOD ·

Use a potato peeler to peel the lemons, making the rind strips as long as possible. Peel at least 6 strips. Wind the strips around the chicken parts, at least one per piece.

Squeeze the juice from the lemons and pour it into a casserole. Add the rest of the vegetables and herbs and finally the chicken. Pour the broth over the top. Cover the pot tightly and put it into a preheated 350°F oven. Cook for 2 hours. Before serving, discard the peel from the chicken pieces.

Serve with the vegetables and cooking liquid.

· SERVING SUGGESTION ·

Rice and pasta (both brown, of course) are the most suitable accompaniments to stewed chicken dishes. Plain boiled whole wheat noodles are fine for every day, but if you are making this dish for a special occasion, serve the brown rice with raisins and pine nuts, as in Braised Chicken with Brown Rice and Raisins (page 108).

INGREDIENTS

4 lemons

6 large chicken parts (leg or breast), skinned

1 large garlic clove

1 onion, stuck with 2 cloves

1 turnip, peeled and quartered

1 leek, sliced in 3 pieces

1 carrot, sliced lengthwise

1 bay leaf

½ teaspoon dried mixed herbs

1 bunch parsley

2 cups chicken broth (page 24)

6 servings

Picture: page 126

Calories	221	★ ★ ★
Fat	7g	★ ★ ★
Saturated Fat	2.4g	★ ★
Cholesterol	150mg	★ ★
Sodium	248mg	★ ★
Fiber	2.1g	★

INGREDIENTS

2-lb roasting chicken

1½ cups low-fat plain yogurt

2 dried chilies, crushed or ground

1 tablespoon paprika

1 tablespoon chili powder

6 garlic cloves, skinned and chopped

5 tablespoons chopped coriander leaves

SQUASH OR ZUCCHINI RAITA

4oz squash or zucchini

1 cup low-fat plain yogurt

1 teaspoon ground cumin

1 teaspoon dried mint leaves

CUCUMBER RAITA

1 long cucumber

¼ cup fresh lemon juice

1 tablespoon chopped coriander leaves

SPINACH RAITA

2 cups spinach

1 cup low-fat plain yogurt

1 teaspoon fenugreek seeds

1 teaspoon chili powder

1 teaspoon ground cumin

NUTTY RAITA

½ cup blanched almonds, halved

½ cup raw cashew nuts

1 teaspoon white cumin seeds

1 teaspoon poppyseeds

2 teaspoons raw sesame seeds

1 cup low-fat plain yogurt

DAIKON AND CARROT RAITA

1 daikon

1 seedless tangerine or orange

2 carrots, grated

1 tablespoon lemon juice

4 tablespoons chopped fresh coriander

6 servings

Picture: page 130

CHICKEN TIKKA WITH FIVE RAITAS

**This cooking method, which is almost a cross between broiling and steaming, is peculiar to the cooking of India.
As usual with chicken recipes from this part of the world, the meat is nice and lean, and always skinned. Raitas are refreshing dips and accompaniments. All the ingredients can be bought in the oriental section of supermarkets.**

· METHOD ·

Make all but the Daikon and Carrot Raita well in advance, as they should be served chilled in order to counteract the hot, spicy flavor of the chicken.

SQUASH or ZUCCHINI RAITA: peel and grate the squash or zucchini. Combine it with the yogurt and the mint leaves. Chill until required.

CUCUMBER RAITA: peel and slice the cucumber as thinly as possible. Put it in a glass or china dish with the lemon juice and chopped coriander.

SPINACH RAITA: wash, trim and chop the spinach finely. Put it into a dry saucepan and cook over medium heat until wilted, about 2 minutes. Leave it to cool, then squeeze out excess water. Whisk the yogurt with the fenugreek, chili powder and ground cumin. Combine with the spinach. Chill until required.

NUTTY RAITA: heat a dry nonstick skillet. Add the almonds, cashews, white cumin, poppyseeds and sesame seeds and fry, stirring occasionally, until they give off an aroma. Combine them with the yogurt.

Skin the chicken, discard all the fat, and ease the flesh from the bones. Cut the chicken into serving pieces and prick the flesh all over with a fork. Mix the yogurt, chilies, paprika, chili powder and garlic in a bowl. Add the chicken and stir well. Cover and refrigerate for 12 hours or overnight, turning occasionally.

Heat a heavy saucepan (not an enamel one) until very hot. Have the lid handy. Sprinkle the coriander leaves over the chicken mixture and empty the contents of the bowl into the saucepan. Quickly cover with the lid. Cook on high for 5 minutes, then reduce the heat and simmer until only a tablespoon or two of the marinade is left, about 20 minutes. Stir the chicken to coat evenly, and serve the chicken hot with all the sauce from the pan.

While the chicken is cooking, prepare the Daikon and Carrot Raita. Shred the daikon and peel the tangerine or orange. Grate the carrots.

Arrange them in a dish and sprinkle them with the lemon juice and the coriander leaves. Chill until required. Serve the chicken on a warmed serving dish and the raitas around it in separate glass bowls.

· SERVING SUGGESTION ·

This delicious dish should be served with any kind of Indian flatbreads (pappadums, pooris, chapatis, naan, etc.), either homemade or reheated before serving. Naan can easily be made at home by flattening a piece of white bread dough (page 29) and broiling or barbecuing it on a very high heat for 10 minutes. Otherwise serve with pita.

Calories	131	★ ★ ★
Fat	4g	★ ★ ★
Saturated Fat	1.4g	★ ★ ★
Cholesterol	70mg	★ ★ ★
Sodium	108mg	★ ★ ★
Fiber	1.1g	★

excluding accompanying raitas

CHICKEN BREASTS WITH GREEN PEPPERCORNS

This is a low-fat variation on the rich, creamy sauces in which chicken breasts are served on special occasions. The recipe also contains no salt.

· METHOD ·

Lay the chicken meat between two sheets of nonstick baking paper. Pound it with a mallet to flatten it. Mix the allspice with the unbleached flour in a shallow bowl. Dip the chicken breasts in the flour, shaking off the excess. Reserve the remaining flour mixture.

Heat the oil in a nonstick skillet over medium heat. Add 4 pieces of chicken at a time; brown for 3 minutes on each side. Remove from the skillet, drain on paper towels and keep warm.

Put the garlic, green onions, lemon juice and wine into the skillet. Add half the peppercorns and cook, stirring to dislodge any bits that have stuck to the bottom of the skillet. Stir 2 teaspoons of the seasoned flour into the yogurt and pour this mixture over the chicken. Cover the skillet and reduce the heat. Simmer on very low heat for 5 minutes, but do not let it boil or the yogurt will separate. Carefully transfer the contents of the skillet to a warmed serving dish and sprinkle the mixture with the reserved green peppercorns.

· SERVING SUGGESTION ·

Attractive whole wheat pasta shapes – such as shells, twists, bowties or rigatoni (ridged macaroni) – will make an appealing accompaniment. You should also serve a green vegetable, such as broccoli.

INGREDIENTS

4 whole chicken breasts, skinned,
   boned and split in half (about 2lb)
½ teaspoon ground allspice
½ cup unbleached flour
2 tablespoons oil
2 garlic cloves, chopped
2 green onions, chopped
Juice of 1 lemon
¾ cup dry white wine
2 teaspoons green peppercorns, crushed
¾ cup low-fat plain yogurt

4 servings

Calories	301	★ ★
Fat	12g	★ ★
Saturated Fat	2.4g	★ ★
Cholesterol	79mg	★ ★
Sodium	116mg	★ ★ ★
Fiber	0.5g	★

INGREDIENTS

3 chicken breasts, skinned and boned

2 cups snow peas

2 cups green beans, sliced in thirds crosswise

2 cups carrots, cut into matchstick strips

3 celery stalks, cut into matchstick strips

3 zucchini, cut into small rounds

1 tablespoon chopped mixed herbs

1 tablespoon salt substitute (page 26)

6 servings

Picture: page 106

Calories	156	★ ★ ★
Fat	3g	★ ★ ★
Saturated Fat	1.1g	★ ★ ★
Cholesterol	69mg	★ ★ ★
Sodium	149mg	★ ★ ★
Fiber	5.5g	★ ★

INGREDIENTS

2 lemons

3 Rock Cornish hens (about 1lb each)

½ teaspoon salt

1 teaspoon black pepper

1 teaspoon ground coriander

1 teaspoon ground cumin

1 teaspoon ground ginger

2 teaspoons chili powder

6 cardamom pods

4 persimmons

3 kiwi fruit

8 servings

Picture: page 122

CHICKEN PACKETS WITH SPRING VEGETABLES

This is the simplest and healthiest dish you can imagine, but the contents of the foil packets must be thoroughly cooked through. Never put too many in your oven, and keep them all in the center. Chicken is tender enough to be ideal for this method, and the vegetables will retain all of their natural goodness.

· METHOD ·

Cut the chicken breasts into bite-sized pieces. Trim the snow peas. Mix all of the vegetables together and sprinkle them with the mixed herbs and salt substitute.

Lay a sheet of foil on a work surface. Cut it into 6-in-wide strips. Cut the strips into pieces 8in long.

Arrange 2 chicken pieces and a tablespoon of vegetables in the center of each foil rectangle. Fold the strip over the top, then fold the edges over several times to seal in the contents. You should have about 20 packets. Put them into a preheated 350°F oven and cook for 30 minutes. Check one of the packets, opening it carefully as a lot of steam will escape. If it is done, you can serve the packets. Let the diners unwrap them for themselves.

· SERVING SUGGESTION ·

New potatoes or potato balls served with lots of parsley are the obvious choice of starch for this springtime dish. A good accompaniment is a platter of mixed baby carrots and peas.

SPICED ROCK CORNISH HENS WITH FRESH FRUIT SAUCE

Uncooked fruit sauces are very popular with desserts. This novel fruit sauce is particularly good with roast meat. Persimmons are in season in midwinter and are very rich in vitamin A in the form of carotene. Since this vitamin is usually associated with fats and oils, it is a good way of absorbing it without eating oil. The Cornish hens can be charcoal-broiled if they are split into serving portions before sprinkling with spices.

· METHOD ·

Squeeze the juice from one of the lemons and sprinkle it over the Rock Cornish hens, inside and out. Mix the salt, pepper, coriander, cumin, ginger and chili powder and sprinkle this mixture over and inside the birds. Crush the cardamom pods and extract the tiny seeds; sprinkle them inside the birds.

Preheat the oven to 400°F. Place the birds side by side on a trivet over a roasting pan (they will taste even better if spit-roasted). Roast them for 20 minutes, then reduce the heat to 350°F and cook for a further 45 minutes, or until well browned.

Squeeze the other lemon. Purée the persimmons (they do not need to be peeled) with two of the peeled kiwi fruit and the lemon juice in a blender or food processor. Warm it very slightly in a saucepan, but do not let it boil.

To serve, cut the Rock Cornish hens in half, place each half on an individual serving dish and pour a little sauce around it. Slice the remaining kiwi fruit and use to decorate each serving.

· SERVING SUGGESTION ·

Serve with lots of fresh watercress, bean sprouts and parsley. Plain brown rice will offset the spiciness of the meat and is a good way to mop up the delicious sauce.

Calories	267	★★
Fat	15g	★★
Saturated Fat	5.9g	★
Cholesterol	124mg	★★
Sodium	243mg	★★
Fiber	1.0g	★

CHICKEN AND MELON SALAD

This refreshing summer salad can be served either as a starter or a light main course, perhaps as a luncheon dish. To make sure the melon is ripe, smell it. If it has a strong perfume, it is ready to eat. Melons are low in calories and have recently been found to contain valuable trace elements.

· METHOD ·

Cut the melon in half, scoop out the seeds and discard them. Scoop out the flesh, leaving a thin layer inside the melon halves, and reserve it. Cut the chicken into bite-sized pieces and combine it with the bell pepper, mayonnaise, mint, white pepper and paprika until well mixed. Pile the mixture into the melon shells and chill until required. Garnish with mint leaves just before serving.

· SERVING SUGGESTION ·

As a starter, this salad needs no accompaniment, but as a main course, a salad of grapefruit and tangerine sections would make a refreshing accompaniment for this summery dish. Crackers or crispbread would round off the meal.

INGREDIENTS

1 small melon (Ogen or Charentais varieties)

8oz cold cooked chicken, with the skin removed

1 green bell pepper, seeded and thinly sliced

6 tablespoons Special Low-fat Mayonnaise (page 26)

1 tablespoon chopped fresh mint

½ teaspoon white pepper

⅛ teaspoon paprika

Mint leaves for garnish

2 servings

Calories	240	★★★
Fat	7g	★★★
Saturated Fat	2.4g	★★
Cholesterol	121mg	★★
Sodium	191mg	★★★
Fiber	2.1g	★

SPICED ROCK CORNISH HENS WITH FRESH FRUIT SAUCE
for recipe see page 120

INGREDIENTS

*2-lb chicken (preferably a boiling fowl),
    cut into serving pieces, skinned,
    all visible fat removed*

½ teaspoon salt

2 small turnips, peeled

2 medium carrots, sliced lengthwise

1 onion, stuck with 2 cloves

1 strip of lemon peel

*1 bunch fresh coriander,
    trimmed*

1 bunch fresh parsley

5 black peppercorns

Juice of 2 lemons

1 teaspoon chili powder

4oz rice noodles

Extra lemons and parsley for serving

6 servings

Picture: page 103

Calories	236	★ ★ ★
Fat	6g	★ ★ ★
Saturated Fat	1.9g	★ ★ ★
Cholesterol	116mg	★ ★
Sodium	266mg	★ ★
Fiber	3.5g	★ ★

Based on 10% extraction in cooking

LEMON CHICKEN SOUP WITH HERBS

**The sign of good chicken soup used to be golden globules of fat
floating on the surface.
It was ceremonially served to Jewish brides and their grooms on
their wedding day in Poland; the fat was supposed to represent
wealth. Nowadays, we know better.
Despite the dire warnings of the old wives, the loss of flavor
caused by removing the fat from soup is imperceptible. To make
sure the soup is made as low-fat as possible, it is advisable to make
it well in advance and chill it in the refrigerator, so that the fat
congeals on the surface and can be completely removed before
serving. This version, rich in vitamin C supplied by the fresh
lemons and parsley, is ideal for cold winters.**

· METHOD ·

Sprinkle the chicken pieces with salt. Pour 2 quarts water into a large
saucepan or casserole with a tight-fitting lid and add the chicken pieces.
Bring the soup to the boil, skimming frequently. When it boils, add the
turnips, carrots and onion and a strip of lemon peel. Simmer for 30
minutes, then add the coriander, parsley, peppercorns, lemon juice and
chili powder. Cover and simmer for another hour.

Strain the soup, reserving the meat, and chill it, preferably overnight.
Remove any fat with a metal spatula. Reheat the soup, coarsely chop the
reserved chicken, and add it to the saucepan. Shred the vegetables and
herbs and stir them into the soup. Add the rice noodles just 2 minutes
before serving. Serve sprinkled with chopped parsley and pass lemon
wedges to be squeezed into the soup.

· SERVING SUGGESTION ·

This is a complete meal in itself. If the rice noodles, which are very thin
and transparent, do not provide sufficient starch for some tastes, serve
plain boiled brown rice on the side.

BARBECUED CHICKEN WITH GRAPES

The small, seedless green grapes used in this recipe look very
pretty in the sauce poured over the chicken and provide
a fresh, contrasting flavor.
Chicken can be difficult to barbecue, since the outside browns
quickly while the inside may still be raw. The way to overcome this
problem is to flatten the chicken before cooking, a technique well
known to the professionals. When you have done it once
you will find it easy.
Chicken, Rock Cornish hen, pigeon or even quail can be
prepared in this way.

· METHOD ·

To flatten the chicken, insert a long, sharp knife along one side of the
backbone and cut the chicken right through, beginning at the neck end.
Then make a similar cut along the other side of the backbone. Pull the
backbone away with the bone that connects the breasts. Discard the
bones, or use them for broth. Pull out the white cartilage in the breast and
discard it. Loosen the skin around the leg and thigh and push it back to
expose the thigh joint. Cut halfway into each thigh joint and pull the legs
straight. Make a small slit below each chicken breast, below the ribs,
right through the skin.

Cover the chicken with nonstick baking paper and pound it with a
mallet to flatten it, taking care not to hit the leg bones. Then twist the legs
inward and up and under the slits below the breasts.

Press the black grapes through a sieve to release the juice and discard
the seeds and skins. Mix the juice with the wine vinegar. Rub the bird
with garlic and sprinkle it with the vinegar and grape juice mixture.

Prepare the barbecue. When the coals are ready, place a weight (such
as a dry skillet containing tin cans or weights) on top of the chicken and
broil it skin side up for 15 minutes. Remove the weights and turn the
chicken over. Baste it with some of the grape juice and cook until the skin
is brown and crisp, about 10 minutes.

Remove the stems from the seedless grapes. Warm the rest of the
grape juice mixture and add the grapes. Serve this as a sauce for the
chicken.

· SERVING SUGGESTION ·

Rice is the natural choice to accompany this dish. First, stir-fry the rice in
a tablespoon of oil, then steam it in a tightly closed pot, putting a double
thickness of cheesecloth over the rice before adding the lid, to absorb the
steam. Cook for 45 minutes. Serve the rice in a large mound, generously
sprinkled with chopped fresh coriander. You can also stir in a few dark-red
kidney beans for a colorful contrast.

INGREDIENTS

2-lb roasting chicken
2 cups black grapes
6 tablespoons white wine vinegar
2 garlic cloves
2 cups Thompson seedless grapes

4 servings

Calories	307	★ ★
Fat	15g	★ ★
Saturated Fat	5.8g	★
Cholesterol	135mg	★ ★
Sodium	109mg	★ ★ ★
Fiber	0.7g	★

LEMONY CHICKEN-IN-THE-POT
for recipe see page 117

INGREDIENTS

*1 whole chicken breast, skinned
and boned*

2 tablespoons cornstarch

1 egg white

2 tablespoons oil

1½ cups walnuts

1 green bell pepper, seeded and diced

1 red bell pepper, seeded and diced

2¼ cups bean sprouts

2 tablespoons hoisin sauce

2 tablespoons chicken broth (page 24)

2 tablespoons dry white wine

4 servings

Picture: page 110

Calories	367	★ ★
Fat	26g	★
Saturated Fat	3.2g	★ ★
Cholesterol	35mg	★ ★ ★
Sodium	157mg	★ ★ ★
Fiber	2.9g	★

CHINESE CHICKEN AND WALNUTS

**This recipe is based on one of those Chinese dishes in which
everything is deep-fried.
Chinese deep-frying is done on high heat, which seals the food
without admitting too much oil. In this recipe the oil has been
reduced and the vegetables combined in an even healthier way,
which probably ends up tasting better than the original!**

· METHOD ·

Cut the chicken breast into 4 pieces. Toss it in the cornstarch, then in the egg white, coating well.

Heat the oil in a wok or nonstick skillet. When it is very hot, add the chicken pieces and sauté them until they are well browned. Remove the chicken pieces and drain them. Replace them with the walnuts, bell peppers and bean sprouts. Stir-fry them briefly until the bean sprouts are transparent but not wilted, about 3 minutes. Drain on paper towels and place in a low oven to keep warm. Put the hoisin sauce into the wok or skillet and add the broth and wine. Cook, stirring, for 1 minute. Slice the chicken pieces into thin strips, put them into the pan, and toss them to cover with sauce.

Arrange the chicken in sauce in the center of a warmed serving dish and arrange the vegetables around it. Serve immediately.

· SERVING SUGGESTION ·

This dish goes very well with whole wheat noodles as a substitute for the classic rice. You could also try soba, Japanese buckwheat noodles, for an interesting change.

INGREDIENTS

*6 large chicken breasts, skinned, all
visible fat removed*

1 teaspoon dried mixed herbs

1 tablespoon Worcestershire sauce

6 cherry tomatoes, peeled and chopped

1 cup tomato juice

2 cups green beans

8oz whole wheat tagliatelle

8oz green tagliatelle

6 servings

Picture: page 135

CHICKEN PASTA PRIMAVERA

**This is a dish of straw-and-hay tagliatelle with spring vegetables of
your choice. The chicken is first sautéed to seal in the juices, and
then slow cooked. Like all chicken dishes, it is relatively low in fat,
and the pasta and vegetables turn it into a complete meal.**

· METHOD ·

Sprinkle the chicken breasts with the mixed herbs and the Worcestershire sauce. Place them in a nonstick skillet and cook over medium heat until the pieces are lightly browned, turning frequently.

Transfer the pieces to a casserole and add the peeled and chopped tomatoes and the tomato juice. Cover and simmer on low heat for 30 minutes. Meanwhile, string the beans and cut them in thirds crosswise. Add the beans to the chicken stew and cook for a further 15 minutes.

Prepare a large pan of boiling water. Add the whole wheat tagliatelle to the pot and cook for 12 minutes or according to package instructions. Remove, drain, rinse and replace them with the green tagliatelle, and cook in the same way. Keep warm.

Arrange the two lots of pasta in two "nests" on a serving dish.

Arrange 3 chicken pieces in each of them and pour the cooking sauce over everything.

· SERVING SUGGESTION ·

This meal is simple to prepare and needs no accompaniment, except perhaps some whole wheat bread to soak up the gravy, which is quite liquid.

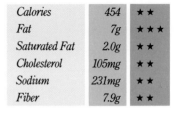

Calories	454	★★
Fat	7g	★★★
Saturated Fat	2.0g	★★
Cholesterol	105mg	★★
Sodium	231mg	★★
Fiber	7.9g	★★

CHICKEN KORMA

Like many East Indian dishes, this one is low in calories. Normally, it is cooked in large amounts of ghee or oil, but only a small amount of oil is actually needed. The spices give such a strong flavor to the food that salt is unnecessary.

· METHOD ·

Wash the chicken pieces and dry them well. Combine the yogurt with half the garlic and smear it over the chicken pieces. Leave them to stand at room temperature for 2 hours, basting occasionally.

In a large, nonstick skillet with a lid, heat the oil and sauté the rest of the garlic with the onions, ginger and cloves. Add the curry powder, garam masala and ground almonds, and continue cooking for another 5 minutes, stirring constantly. Cover the pan and simmer for 2 hours, stirring occasionally.

· SERVING SUGGESTION ·

Long-grain brown rice is the natural accompaniment to this classic East Indian dish. A refreshing cucumber raita (see Chicken Tikka with Five Raitas, page 118) or a simple garnish of watercress would make a suitable vegetable accompaniment.

INGREDIENTS

1 large chicken (about 5lb), disjointed and skinned

2 cups low-fat yogurt

4 garlic cloves, chopped

1 tablespoon oil

2 medium onions, chopped

½ teaspoon ground ginger

½ teaspoon ground cloves

1 tablespoon curry powder

1 tablespoon garam masala

1 teaspoon ground almonds

6 servings

Calories	272	★★
Fat	11g	★★
Saturated Fat	3.0g	★★
Cholesterol	138mg	★★
Sodium	203mg	★★
Fiber	1.1g	★

CHICKEN TIKKA WITH FIVE RAITAS
for recipe see page 118

INGREDIENTS

1 tablespoon oil

1 medium onion, finely chopped

3½-lb chicken, cut into serving pieces, skinned, all visible fat removed

½ teaspoon ground turmeric

½ teaspoon pepper

Strained juice of 3 pomegranates

1 ¾ cups walnuts, coarsely chopped

1 cup fresh orange juice

4 servings

Calories	519	★
Fat	34g	★
Saturated Fat	4.5g	★ ★
Cholesterol	149mg	★ ★
Sodium	142mg	★ ★ ★
Fiber	2.7g	★

INGREDIENTS

2-lb roasting chicken, skinned, all visible fat removed

2 small onions, 1 stuck with 2 cloves

1 small carrot

3 slices stale white bread, crusts removed

3¾ cups walnuts

2 teaspoons paprika

¼ teaspoon salt or salt substitute (page 26)

6 servings

FASENJAN CHICKEN

Braising the breasts in this delicious pomegranate-and-walnut sauce is a great low-fat way to cook chicken and exotic enough to serve at a special meal.
The sauce can also be used to cook duck or lamb.

· METHOD ·

Heat the oil in a nonstick skillet with a lid. Add the onion and sauté until it is transparent. Then add the chicken pieces and sprinkle them with the turmeric and pepper. Brown them lightly, turning frequently to cook evenly. Remove the chicken pieces from the skillet and place them on paper towels to drain.

Add the walnuts to the skillet and cook, stirring, for 2 minutes. Then add the pomegranate juice, orange juice and 1 cup warm water and bring to the boil.

Return the chicken pieces to the skillet, cover the skillet and simmer over low heat for 45 minutes. Serve hot.

· SERVING SUGGESTION ·

The chicken should be served with plain boiled brown rice, or with Festival Pilaf (page 72), cooked without the ground meat. A green salad of lettuce, celery, coriander leaves and parsley will go well with this dish.

CIRCASSIAN CHICKEN

This chicken-and-walnut dish is not hard to make but it is festive enough for a buffet or party. It also makes an excellent appetizer for a dinner party, if arranged in individual dishes.
It is low in fat and salt, despite being a totally authentic, traditional Turkish recipe. The oil in walnuts, as well as having a delicious flavor, is high in polyunsaturates.

· METHOD ·

Put the chicken into a large pot, together with the onion stuck with cloves and the carrot.

Add 7 cups of water and bring it to the boil, skimming off any scum that rises to the surface. Cover the pot and let the chicken simmer, covered, until cooked through, about 1½ hours.

Remove the chicken, reserving the liquid. Let the chicken cool, then bone it and cut the flesh into 1-in pieces. Place them in a bowl.

While the chicken is cooking, soak the bread in a little of the reserved chicken broth. Squeeze it dry. Mince the remaining onion. Grind the walnuts finely in a food processor. With the machine running, add the squeezed bread, the minced onion, paprika and salt. Transfer the mixture to a sieve over a bowl, to allow the walnut oil to drain out of it. It should drain for at least 30 minutes. Reserve the drained oil.

Transfer the walnut mixture in the sieve to a casserole and add 4 cups chicken broth. Stir well, to make a thick sauce. Mix half the sauce thus obtained in the bowl with the chicken pieces. Arrange the mixture on a large flat serving platter and pour the rest of the sauce over the top. Garnish with the olives and cucumber. Sprinkle a little of the reserved walnut oil over the dish.

· SERVING SUGGESTION ·

Serve with a salad consisting of chopped cucumber, bell peppers, tomatoes and shredded lettuce. Sprinkle it with a little more of the walnut oil. Eat the chicken by scooping it up on pita.

Calories	526	★
Fat	38g	★
Saturated Fat	5.2g	★
Cholesterol	116mg	★ ★
Sodium	195mg	★ ★ ★
Fiber	4.0g	★ ★

BAKED CHICKEN WITH PRUNES

This is a popular way to cook chicken in eastern Europe and in Scandinavia, though the idea probably comes from the Middle East, where medieval cooks were combining fruit and meat long before the European crusaders arrived to steal the recipes! Prunes are rich in iron, but chicken is not, and the two flavors complement each other nicely.

· METHOD ·

Heat the oil in a nonstick skillet and add the chicken pieces. Brown lightly.

Transfer the chicken pieces to a Dutch oven. Add the vegetables, bay leaf, parsley and salt, and enough water to just cover the meat. Place the lid on the pot and stew the contents for 45 minutes.

Drain the prunes and put them into a pot with the lemon juice and 1 cup water. Stew them, covered, on low heat for 20 minutes.

Remove the chicken pieces from the cooking liquid and arrange them on a serving dish. Remove the prunes with a spoon and arrange them around the chicken. Strain the chicken cooking liquid and pour 1 cup into a small bowl. Let it cool slightly, then whisk in the cornstarch. Pour the prune cooking liquid into the chicken and cornstarch mixture and cook on low heat, stirring frequently, until the sauce thickens. Pour some over the chicken pieces and serve the rest separately.

· SERVING SUGGESTION ·

Boiled potatoes are the best accompaniment to this chicken dish, and boiled or steamed carrots will make an attractive color contrast to the prunes. Sprinkle the cooked vegetables with chopped fresh parsley.

INGREDIENTS

1 tablespoon oil

4-½lb chicken, cut into serving pieces, skinned, all visible fat removed

1 carrot, split lengthwise

1 celery stalk

1 onion, stuck with 2 cloves

1 bay leaf

1 sprig parsley

½ teaspoon salt

1½ cups pitted prunes, soaked in tea for 2 hours

1 tablespoon lemon juice

1 tablespoon cornstarch

6 servings

Picture: page 134

Calories	271	★ ★
Fat	9g	★ ★
Saturated Fat	2.4g	★ ★
Cholesterol	132mg	★ ★
Sodium	124mg	★ ★ ★
Fiber	6.7g	★ ★

BAKED CHICKEN WITH PRUNES

for recipe see page 133

CHICKEN PASTA PRIMAVERA

for recipe see page 128

INGREDIENTS

2 cooked chicken breasts, boned

1 tablespoon oil

4 shallots, finely chopped

4 garlic cloves, finely chopped

*1lb cooked shrimp, shelled
and deveined*

*1 tablespoon Asian fish sauce or
Worcestershire sauce*

2 teaspoons sugar

Juice of 2 limes

⅓ cup dry-roasted peanuts, chopped

8 lettuce leaves

*2 tablespoons chopped fresh
coriander*

¼ cup sesame seeds

8 servings

Picture: page 138

Calories	208	★ ★ ★
Fat	11g	★ ★
Saturated Fat	1.9g	★ ★ ★
Cholesterol	133mg	★ ★
Sodium	996mg	★
Fiber	1.3g	★

INGREDIENTS

¾ cup dry whole wheat breadcrumbs

1 teaspoon dried basil

1 teaspoon dried tarragon

½ teaspoon onion powder

½ teaspoon ground black pepper

1 egg, beaten

Juice of 1 lemon

*12 chicken drumsticks, skinned
(about 8oz each)*

*2 zucchini, sliced crosswise into
1-in rounds*

1 tablespoon oil

6 servings

CHINESE CHICKEN SALAD WITH SESAME SEEDS

Chicken and shrimp go well together, and both are low in fat, although shrimp have more cholesterol. In order to get the maximum benefit from the vitamins in the raw vegetables, they should be prepared just before eating, although the meat and fish can be cooked well ahead of time. Use an oil high in polyunsaturates, such as safflower or grapeseed, for the frying.

· METHOD ·

Cut the chicken into bite-sized pieces. Heat the oil in a wok or skillet and stir-fry the shallots and garlic until transparent. Drain well on paper towels.

Combine the chicken pieces and the shrimp. Mix the fish sauce or Worcestershire sauce with the sugar and lime juice. Pour this mixture over the chicken and shrimp, and sprinkle with the peanuts. Stir in half the shallots and garlic.

Arrange the chicken-and-shrimp mixture on a bed of lettuce leaves. Sprinkle it with the rest of the shallots and garlic, and finally decorate with the sesame seeds.

· SERVING SUGGESTION ·

If this dish is served as an appetizer, there is really nothing to add to it. However, it can also form part of a Chinese-style main course, in which case whole wheat noodles steamed with Chinese vegetables such as snow peas, baby corn, bean sprouts and bamboo shoots would be the perfect accompaniment.

BAKED HERBED CHICKEN DRUMSTICKS

For people who like chicken with a crisp coating, Southern-style, you can still have this kind of dish without having to resort to deep-fat frying, and the recipe contains less salt than the classic version. You can experiment with similar coatings by using rolled oats or matzo meal instead of breadcrumbs.

· METHOD ·

Combine the crumbs, herbs and spices in a shallow bowl and beat the egg and lemon juice in another shallow bowl. Pat the chicken and zucchini with paper towels to make sure they are dry. Coat them first in the egg mixture, then in the breadcrumb mixture, making sure they are evenly and thoroughly coated. Heat the oven to 350°F.

Put a flameproof baking dish on the stove; it should be large enough to hold the drumsticks in a single layer. Pour the oil into it, and swirl it around to coat the bottom and sides. Leave it on medium heat for 2 minutes, then add the drumsticks. Put the dish into the oven and bake, turning every 15 minutes, for 1 hour, or until the coating is crisp and the chicken cooked through.

· SERVING SUGGESTION ·

This is a wonderful picnic treat, snack meal, or buffet dish to serve with drinks. If eaten in this way, it would be delicious with raw vegetables (*crudités*) and a yogurt or Special Low-Fat Mayonnaise (see page 26) dip. If eaten at home, serve it with mashed potatoes and brussels sprouts.

Calories	373	★ ★
Fat	16g	★
Saturated Fat	5.0g	★ ★
Cholesterol	280mg	★
Sodium	359mg	★ ★
Fiber	2.4g	★

CHINESE CHICKEN WINGS

Wings are very inexpensive chicken parts, but they cannot be skinned easily, so they should be cooked without extra fat. They make a delicious salad dish cooked in this way. Worcestershire sauce has been substituted for Asian fish sauce, which can be used if available.

INGREDIENTS

24 chicken wings

2 tablespoons soy sauce

2 tablespoons Worcestershire sauce

2 tablespoons brown sugar

1 tablespoon dry sherry

2 very thin pieces ginger root, chopped

1 whole star anise or ¼ teaspoon fennel seeds

6 servings

· METHOD ·

Rinse and drain the chicken wings and inspect them for any remaining feathers or hairs, which can be pulled out or singed off. Put them into a shallow pan with a lid, and add the rest of the ingredients. Pour in 6 tablespoons water and bring to the boil. Cover the pan, reduce the heat and simmer for 20 minutes, stirring at least twice to keep the wings from sticking to the bottom of the pan. Uncover the pan and cook for another 15 minutes, basting the wings, until only about ¼ cup of liquid remains and the wings are an attractive dark color. Cool, then refrigerate them until required.

· SERVING SUGGESTION ·

Slice a Chinese cabbage crosswise into slices about ½in thick. Arrange a slice on each of 6 salad plates and arrange 4 wings on each plate. Pour a little of the sauce over them. Serve as a starter, or as a main course with plain boiled brown rice.

Calories	340	★ ★
Fat	22g	★
Saturated Fat	8.7g	★
Cholesterol	136mg	★ ★
Sodium	153mg	★ ★ ★
Fiber	0.0g	★

CHINESE CHICKEN SALAD WITH SESAME SEEDS

for recipe see page 136

TURKEY AND OTHER FOWL

Turkey and other fowl are sold fresh or frozen. It is always preferable to buy fresh birds; for one thing, you can be sure that you are getting the whole weight of the bird, not paying for frozen water. Also, it is much safer to cook a fresh bird rather than a thawed one. If you buy a frozen bird, make sure you allow plenty of time for thawing – at least 12 hours at room temperature for a 9lb turkey, for instance – or defrost it in a microwave oven. The manufacturer will provide timings for defrosting.

As with chicken, try to find birds that have their giblets; the giblets can be used to flavor soup and roasting juices.

All fowl need thorough cooking, except duck, for which the breast can be left pink. To check whether the meat is cooked, try moving the leg. If it moves easily and the juices flowing from it are clear, the bird is cooked through.

A standard portion is 6oz meat per person and 9oz if the meat is on the bone.

DEVILED TURKEY LEGS WITH PURÉED VEGETABLES

The skin of the legs does not have a layer of fat beneath it as does the skin on the breast and so is already lower in calories and fat. But it can be discarded for weight-watchers, as the flavoring will have penetrated into the flesh.
Technically speaking, deviled foods should be grilled. However, the thickness of the turkey bone makes this a rather uncertain operation, so it is advisable to roast the meat in a hot oven.
The blandness and texture of puréed vegetables serve as a complementary contrast to the hot, spicy flavor of the meat, as well as providing the carbohydrate for a balanced meal.

· METHOD ·

Mix the mustard, chili powder and herbs until well combined. Gradually mix in the Worcestershire sauce and the Tabasco sauce. Use as much of the lemon juice as is needed to achieve a smooth paste. Preheat the oven to 450°F.

Use a sharp knife to slash the turkey legs all over, crisscrossing the slashes and making sure that they penetrate beneath the skin. Spread the coating mixture on a spatula and rub it into the meat. Do not use your bare fingers, as the mixture will sting. Arrange the legs in a roasting pan just large enough to hold them.

Roast the meat for 20 minutes, then turn it over and roast for another 20 minutes. Remove the legs from the oven and let them cool slightly before removing the meat from the bone (turkey legs have long tendons that make them difficult to eat elegantly at table; also one leg may be too large a portion for one person).

While the meat is cooking, put a large pan of cold water on the stove. Add the potatoes, rutabagas and carrots and simmer them, uncovered, on a fairly low heat for 30 minutes. Drain them, reserving the liquid. Cook the spinach in the liquid; bring it to a brisk boil and cook it for 3 minutes only. Drain well.

Mash the potatoes, combining them with some of the cooking liquid, and season them with the lemon rind, salt or salt substitute, and a pinch of nutmeg. Mash the rutabagas separately, seasoning them with salt or salt substitute, and sprinkling the top with a little parsley. Mash the carrots and season them with the ground coriander. Chop the spinach finely and season it with another pinch of nutmeg.

· SERVING SUGGESTION ·

This makes a very good buffet dish, with the legs in one dish and the vegetables served separately, each in a glass bowl to show off the color contrasts. You will be surprised at how labor-saving it is to cook vegetables in this way. At the same time you will hardly notice the absence of fat in the vegetables. If you wish, however, you can serve some low-fat magarine or yogurt separately.

INGREDIENTS

2 teaspoons dry mustard
2 teaspoons chili powder
1 teaspoon mixed dried herbs
2 teaspoons Worcestershire sauce
5 drops Tabasco sauce
Juice and rind of ½ lemon
4 turkey legs
1lb potatoes
1lb rutabaga
1lb carrots
1lb spinach
½ teaspoon salt or salt substitute (page 26)
½ teaspoon ground nutmeg
Chopped parsley
¼ teaspoon ground coriander

6 servings

Calories	215	★ ★ ★
Fat	5g	★ ★ ★
Saturated Fat	1.5g	★ ★ ★
Cholesterol	97mg	★ ★
Sodium	194mg	★ ★ ★
Fiber	3.6g	★ ★

CLASSIC ROAST TURKEY WITH THREE STUFFINGS
for recipe see page 144

INGREDIENTS

11-lb fresh turkey with giblets

8 Canadian-style bacon strips

1 onion

CHESTNUT STUFFING

8oz Italian dried chestnuts

½ teaspoon ground cloves

1 cup milk

PRUNE STUFFING

24 pitted prunes, soaked in tea overnight

1½ cups seedless raisins, soaked in ¼ cup port overnight

1 turkey liver, cooked and chopped

GROUND ALMOND STUFFING

1 cup ground almonds

4 tablespoons chopped fresh parsley

½ teaspoon mixed dried herbs

4 tablespoons chopped fresh coriander

1 egg

12 servings

Picture: page 142

CLASSIC ROAST TURKEY WITH THREE STUFFINGS

My mother always roasts turkey this way, and it is delicious. The bacon should be discarded by anyone who is on a low-fat, low-salt diet, but the flavor permeates the skin while cooking and makes it taste wonderful. It is most important to roast a turkey slowly, particularly a stuffed turkey, as the heat must penetrate to the center of the bird. A bird that is not thoroughly cooked is unhealthy, as well as being unappetizing.
If the cavity is not big enough for all three stuffings, some or all can be cooked separately.
Always remove any leftover stuffing from inside the turkey *immediately* after it has been served, and store it separately. Otherwise both bird and stuffing could cause food poisoning. For the same reason it is advisable to buy fresh, not frozen, birds.

· METHOD ·

Make the stuffings the night before cooking the turkey, but do *not* stuff the bird until you are just about to cook it. Soak the chestnuts in the milk for 3 hours, then simmer them gently, adding the ground cloves, until all the liquid is absorbed. Remove from the heat, then grind the chestnuts in a food processor.

To make the prune stuffing, drain the prunes and put them, together with the raisins in their soaking liquid, into a saucepan. Simmer for 15 minutes, then drain. Blend in a food processor with the turkey liver.

To make the almond stuffing, simply mix the almonds and herbs together and bind them with the egg.

Stuff the turkey first with the prune mixture, which should go right at the back of the bird. Then stuff with the chestnut stuffing, and finally add the almond stuffing.

Preheat the oven to 325°F. Truss the bird firmly, holding the bacon strips well in place. Place the bird on its side in a roasting pan, and put it on the center shelf of the oven. Cook for 1 hour. Then turn the turkey onto the other side and cook for another hour. Now turn breast side up and cover loosely with a piece of foil. Cook for 18 minutes per pound of meat. For the last 20 minutes of cooking remove the foil, and increase the heat to 400°F. The bird is cooked if the juices run clear when the flesh is pierced at the thigh joint, and the legs move easily.

Let the bird rest for 20 minutes before it is carved. Discard the bacon before serving.

· SERVING SUGGESTION ·

Cranberry sauce is the traditional accompaniment but the purchased kind is often full of sugar. Instead, why not purée other fresh or frozen berries, such as blackberries or blueberries, without sugar, and serve them with a little lemon juice?

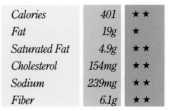

Steamed or microwaved green vegetables, such as peas or brussels sprouts, are very good with turkey. You can roast parboiled sweet potatoes or turnips for the last 30 minutes of the turkey cooking time, in the roasting pan with the bird or separately, wrapped in foil. An alternative is baked potatoes roasted in the oven during the last hour of cooking.

Calories	401	★ ★
Fat	19g	★
Saturated Fat	4.9g	★ ★
Cholesterol	154mg	★ ★
Sodium	239mg	★ ★
Fiber	6.1g	★ ★

TURKEY MONTMORENCY

The classic dish Duck Montmorency is simply duck in cherry sauce. Since turkey is so much lower in fat and has some of the attributes of a game bird, it is about time duck and other game recipes were adapted for turkey. Other turkey parts can be used instead of breasts; the weight given is for boned meat. This is also a great way to use up leftover turkey; if you do, omit the shallow-frying procedure and start from where the cherries are added. You can use canned cherries; in which case omit the honey.

· METHOD ·

Slice the breasts into serving portions. Put the bacon into a nonstick skillet and heat it until the fat begins to run.

Add half the breast slices and brown them. Remove and add the rest of the breast slices. Drain and reserve. Chop the lean parts of the bacon, and discard any visible fat.

Transfer the turkey pieces to a casserole and sprinkle with the bits of bacon. Add the wine, cherries, cherry brandy, broth and cinnamon stick. Cover and simmer for 15 minutes. Dissolve the cornstarch in 3 tablespoons of water and add this to the pot to thicken the sauce.

Transfer the turkey to a warmed serving dish, pour the cherry sauce over it, and sprinkle it with the almonds.

· SERVING SUGGESTION ·

A green vegetable, such as cabbage, curly kale or broccoli, will make a nice color contrast when served with this dish. Brown rice would round off the menu, or you could serve attractive whole wheat pasta shapes, such as shells or wheels.

INGREDIENTS

6 slices skinless turkey breast (6oz each)

2 Canadian-style bacon strips, rinds removed

1 cup dry red wine

1lb sweet cherries, pitted or cherries canned in their own juice

1/4 cup cherry brandy

6 tablespoons chicken broth (page 24)

1 cinnamon stick

1 tablespoon honey (optional)

1 tablespoon cornstarch

1/4 cup blanched slivered almonds

6 servings

Picture: page 147

Calories	338	★ ★
Fat	7g	★ ★ ★
Saturated Fat	1.2g	★ ★ ★
Cholesterol	89mg	★ ★
Sodium	225mg	★ ★
Fiber	2.3g	★

TURKEY SALAD LYONNAISE
for recipe see page 152

TURKEY MONTMORENCY
for recipe see page 145

INGREDIENTS

*4½lb turkey parts with bones (or
    2lb boneless turkey parts)*

1 onion, stuck with 2 cloves

2 celery stalks

3 sprigs parsley

1 carrot, split lengthwise

1lb brussels sprouts

1½lb sweet potatoes

8 servings

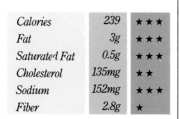

Calories	*239*	★ ★ ★
Fat	*3g*	★ ★ ★
Saturated Fat	*0.5g*	★ ★ ★
Cholesterol	*135mg*	★ ★
Sodium	*152mg*	★ ★ ★
Fiber	*2.8g*	★

BRAISED TURKEY WITH SWEET POTATOES AND BRUSSELS SPROUTS

**This is an excellent way of using turkey if you do not want to cook a whole bird, and the cooking juices can be used for broth if you do not want to serve it as a gravy.
Cooked sweet potatoes have a flavor very similar to that of chestnuts, but are more digestible and much cheaper. Try to use the yellow variety, as they will look more attractive in this recipe.**

· METHOD ·

Skin the turkey pieces and remove all visible fat. Put them into a pot and cover with water. Add the onion, celery, parsley and carrot. Bring to the boil, skimming off any scum. When the liquid boils, reduce the heat and cover the pan. Cook until the turkey is tender, about 2½ hours.

Prepare the brussels sprouts by cutting a cross in the base of each with a sharp knife to allow the sprouts to cook evenly throughout. Bring a large pan of water to the boil and add the sprouts. Boil on moderate heat, uncovered, for 15 minutes, or until tender. Remove the brussels sprouts with a large slotted spoon and reserve.

Reduce the heat and add the sweet potatoes. Cover the pan and simmer them for 30 minutes, or until cooked through – pierce the largest one with a sharp knife to test it. Drain them (you can reserve the water for vegetable broth). When the potatoes are cool enough to handle, peel them with your fingers and a sharp knife and cut them into 2-in cubes. Combine with the brussels sprouts and keep warm in a serving dish.

To serve, bone the turkey pieces and keep them warm while you quickly reheat the vegetables. Arrange the turkey and vegetables side by side in a large serving dish. Serve hot.

· SERVING SUGGESTION ·

The best accompaniment for this dish is mashed potatoes with hot gravy. Alternatively, other root vegetables, such as turnips or rutabagas are most attractive when puréed, or boiled parsnips could be sliced into matchstick strips and served.

TURKEY TONNATO

**This cold turkey dish is a lower-calorie and less expensive version
of the classic Italian *vitello tonnato*.
It is an easy way to ensure that the turkey breast remains moist
and has a good flavor. When the cooking liquid is strained and
chilled, it can be skimmed of fat and makes a good broth.**

· METHOD ·

Rinse the anchovies under running water. With a sharp knife cut 8 slits in
the top of the turkey breast, being sure to penetrate the skin. Cut each
garlic clove in half lengthwise, and 2 of the anchovies crosswise into 4
pieces. Insert a piece of garlic and slice of anchovy securely into each slit.

Put the turkey breast, skin side down, into a large casserole. Add the
onion, carrot, parsley, bay leaf and thyme. Then add the chicken broth
and wine. Bring to the boil, skimming as necessary. Cover the casserole
and simmer over gentle heat for 30 minutes. Then turn the breast over
and simmer it for another 40 minutes. Remove the casserole from the
heat and cool the contents to room temperature.

Remove the turkey breast from the casserole, skin it and ease each
side of the breast off the bone in one piece. Slice the breast neatly into
slices about 1/4in wide and arrange them slightly overlapping on a serving
dish.

In a food processor or blender, grind the tuna and add the Special Low-
Fat Mayonnaise while the machine is running. Spoon this sauce over the
turkey slices. Decorate with the chopped parsley. Rinse the capers to
remove excess salt and dot them over the mayonnaise.

· SERVING SUGGESTION ·

Use fresh salad ingredients or twists of lemon and cucumber to brighten
up this turkey dish and make it fit for the most special of occasions. It
should be accompanied by a big bowl of potato salad, made by combining
boiled and cubed waxy potatoes with more of the Special Low-Fat
Mayonnaise (page 26) and some pieces of chopped pickled gherkin. A
lettuce and cucumber salad completes the picture.

INGREDIENTS

2-oz can anchovies

4-lb turkey breast, with bones

4 garlic cloves, peeled

1 onion, quartered

1 carrot, sliced

4 sprigs parsley

1 bay leaf

1 teaspoon dried thyme

5 cups chicken broth (page 24)

2½ cups dry white wine

*1½ cups Special Low-Fat
    Mayonnaise (page 26)*

7-oz can tuna, packed in water

6 tablespoons chopped parsley

2 tablespoons capers

10 servings

Calories	241	★ ★ ★
Fat	3g	★ ★ ★
Saturated Fat	1.0g	★ ★ ★
Cholesterol	93mg	★ ★
Sodium	424mg	★
Fiber	1.5g	★

ROAST GOOSE WITH FRESH PINEAPPLE AND ALMONDS
for recipe see page 153

INGREDIENTS

4½-lb duck, trimmed of all fat,
  giblets reserved

4 green onions, sliced lengthwise
  toward the white part

4 thin slices ginger root

1 teaspoon Chinese powdered five-spice

3 tablespoons dry sherry

4 tablespoons soy sauce

2 tablespoons dark brown sugar

1 tablespoon oil

6 servings

Picture: page 155

Calories	596	★
Fat	52g	★
Saturated Fat	13.9g	★
Cholesterol	174mg	★ ★
Sodium	139mg	★ ★ ★
Fiber	0.1g	★

INGREDIENTS

12oz cooked skinned turkey meat,
  diced

1 cup flaked smoked whitefish,
  bones carefully removed

4 sour pickles, sliced lengthwise
  and diced

½ cup Special Low-fat
  Mayonnaise (page 26)

6 tablespoons dry white wine

2 tablespoons chopped parsley

1 medium carrot

4 medium potatoes

4 servings

Picture: page 146

Calories	364	★ ★
Fat	10g	★ ★
Saturated Fat	2.6g	★ ★
Cholesterol	114mg	★ ★
Sodium	175mg	★ ★ ★
Fiber	3.5g	★ ★

CHINESE-STYLE SPICED BRAISED DUCK

**This is a variation on the Chinese method of boiling duck.
Normally, the cooking liquid from the first boiling would not be
discarded and replaced, but since it contains most of the fat, it is
preferable from a health point of view to do so.
Very little flavor is lost in the process.**

· METHOD ·

Bring 4 cups water to the boil. Add the duck and boil for 5 minutes.
Remove the duck and pat it dry with paper towels. Discard the liquid.
Remove the skin and surplus fat from the duck.

Bring another 4 cups water to the boil. Add the green onions, ginger
and five-spice to the water. Then stir in the sherry, soy sauce and sugar.
Add the duck and reduce the heat. Cover the pot tightly and cook the duck
at a bare simmer for 2 hours.

Bring the cooking liquid to the boil, uncovered and without stirring,
and keep it on high heat to reduce it. Meanwhile, cut the duck into small
pieces. Arrange the duck on a platter. Coat it with the reduced sauce.

· SERVING SUGGESTION ·

Garnish the duck with green onions, split lengthwise, and slices of carrot.
Serve with brown rice and a salad containing pickles and shredded lettuce.

TURKEY SALAD LYONNAISE

**This delicious and very unusual salad is low in fat and oil.
The combination of smoked fish and turkey is one that can be
explored using other fish.**

· METHOD ·

In a salad bowl, mix the turkey meat, flaked whitefish, pickles, Special
Low-Fat Mayonnaise, parsley and white wine. Refrigerate for 1 hour.
Divide into individual portions and sprinkle with the wine and parsley.
Peel carrot slivers from the carrot with a potato peeler and reserve.

Boil the potatoes in their skins for 20 minutes. Drain, peel and slice
them. Serve hot with the fish salad and decorate with the carrot slivers.

· SERVING SUGGESTION ·

A fresh green salad would make a good accompaniment for this cooked
salad. Cooked beets can be added if you like. Turkey Salad Lyonnaise is an
ideal dish for a formal picnic or outdoor meal.

ROAST GOOSE WITH FRESH PINEAPPLE AND ALMONDS

Goose is very rich, fatty meat and is best left to special occasions. However, this fruity stuffing will counterbalance the richness, and it makes a change from the traditional apple accompaniment. Goose can also be tough (although nowadays it is always killed for the table when young) and fresh pineapple acts as a natural tenderizer. Those who are watching their weight should remove the skin from serving portions.

INGREDIENTS

9-lb goose
1 small pineapple
2 cooking apples
Juice and grated rind of 1 orange
1 cup unsweetened pineapple juice
¼ teaspoon cinnamon
½ teaspoon ground ginger
1 tablespoon honey
6 tablespoons sweet sherry
2 tablespoons brandy

6 servings

Picture: page 150

· METHOD ·

Remove as much external fat as possible from the goose and prick the skin to allow the fat to drain out from under the skin during cooking. Cut the rough skin from the pineapple, remove the eyes and slice it into rings. Cut out the core and slice the rings into pieces. Core the apples, but do not peel them, and chop into pieces. Preheat the oven to 400°F.

Combine the pineapple and apple with half the orange rind, the cinnamon and ¼ teaspoon of the ginger. Stuff this mixture into the bird's cavity. Place the goose on a trivet over a roasting pan and roast it in the oven for 30 minutes. While it is cooking, mix together the pineapple juice, honey, remaining ginger and sherry. Use this mixture to baste the bird, at the same time reducing the heat to 350°F. Cook for a further 1½ hours (or 20 minutes per pound), basting with the cooking liquid every 20 minutes.

Carve the goose into serving pieces and arrange it on a platter. Keep it warm in a low oven or on a hot tray. Pour the cooking juices into a bowl and set it over ice cubes. Skim off the fat with a spoon or paper towels. Add the rest of the orange rind and the brandy. Boil the sauce for 5 minutes to reduce it, then pour it over the goose.

· SERVING SUGGESTION ·

Goose is delicious served with a whole grain such as cracked wheat, whole wheat kernels or barley. The grain must be soaked in plenty of water for at least 1 hour, and can then be simmered, covered, in the oven for the last 1½ hours of cooking, when the heat has been reduced.

Calories	929	★
Fat	58g	★
Saturated Fat	13.8g	★
Cholesterol	237mg	★
Sodium	395mg	★ ★
Fiber	2.2g	★

BRAISED GUINEA FOWL WITH MUSHROOMS AND BLUEBERRIES

for recipe see page 157

CHINESE-STYLE SPICED BRAISED DUCK
for recipe see page 152

GAME AND WILDFOWL

In the US, all the game sold commercially is hand reared, so the dangers inherent in preparing wild birds and animals for the table have largely been eliminated. However carefully the meat has been hung, drawn and plucked, always inspect wild game for shot pellets that may have been left in the body. Some small species of wild duck are not drawn before cooking, but all other game birds need very careful preparation. Pheasant, partridge and grouse need to be hung so the flavor can develop; other wildfowl, especially wild duck, must be eaten within 24 hours of being shot. Wild duck have waterproofing glands near the tail that must be cut out and discarded or they will make the flesh taste bitter. The meat should be rubbed inside and out with a cut onion and a cut lemon before cooking, to counteract its slightly fishy taste.

Furred game meat should be glossy and juicy looking; venison should be dark red, and rabbit pink. Venison should be hung for between 10 and 20 days before eating; rabbit should generally be eaten immediately. Allow one pigeon or small game bird, and half a rabbit or 8oz game animal meat per person.

BRAISED GUINEA FOWL WITH MUSHROOMS AND BLUEBERRIES

Guinea fowl is now domesticated, but the flesh is still very lean and there is always the danger of its drying out, unless it is kept very moist during cooking. For this reason, it is a good idea to cook it in a clay pot (see Clay Pot Chicken with Olives for the cooking method) or in a microwave or combination oven.
The method given here is for a conventional oven, but could easily be adapted by following the manufacturer's instructions. If you cannot find guinea fowl, wild duck or even turkey can be substituted.

· METHOD ·

Remove all fat from the guinea fowl pieces. Pour 1 tablespoon of the oil into a nonstick skillet and heat it. Quickly sauté the guinea fowl over high heat until the meat is no longer red. Transfer the meat to a casserole with a tight-fitting lid.

Pour the second tablespoon of oil into the skillet and add the onions. Fry them until they are transparent, then add the carrots. Cook for another 5 minutes, then toss in the mushrooms. When the mushrooms start to give off their juice, transfer all the ingredients in the skillet to the casserole and add the remaining ingredients. Cover the casserole tightly.

Preheat the oven to 325°F. Put the casserole in the oven and cook for 2 hours. To serve, remove the guinea fowl from the casserole, drain it and season just before serving. Strain the liquid in the pan and pour a little over the meat. Serve the rest in a sauce boat.

· SERVING SUGGESTION ·

Baked potatoes in their jackets or a purée of root vegetables would go well with the guinea fowl. As it will probably be served on a special occasion, a three-color mold could be made of a layer each of boiled and puréed yellow turnips, of puréed green peas and of yams or parsnips. It can be simmered in a water bath in the oven for the last hour of cooking and unmolded just before serving.

INGREDIENTS

1 young guinea fowl, cut into serving pieces
2 tablespoons oil
2 onions, sliced
2 carrots, sliced
1 cup large mushrooms, sliced
4 juniper berries
1 cup blueberries, fresh or frozen
2 tablespoons dry red wine
1½ cups chicken broth (page 24)
1 bay leaf
½ teaspoon black pepper
½ teaspoon salt or salt substitute (page 26)
Extra berries for decoration

4 servings
Picture: page 154

Calories	293	★ ★
Fat	16g	★
Saturated Fat	2.7g	★ ★
Cholesterol	47mg	★ ★ ★
Sodium	193mg	★ ★ ★
Fiber	4.6g	★ ★

ROAST PHEASANT WITH BLACKBERRY AND APPLE

for recipe see page 161

INGREDIENTS

1 haunch of venison (3lb)

2¼ cups fresh or 6oz dried wild mushrooms, soaked in water for 20 minutes

1 teaspoon salt

8 juniper berries, crushed

4 green peppercorns

FOR THE MARINADE

2 onions, sliced

2 carrots, thinly sliced

2 shallots, chopped

1 celery stalk, sliced

1 bouquet garni

4 cups dry red wine

2 tablespoons cornstarch

1 cup low-fat plain yogurt, at room temperature

8 servings

Calories	*396*	★★
Fat	*10g*	★★
Saturated Fat	*0.3g*	★★★
Cholesterol	*111mg*	★★
Sodium	*312mg*	★★
Fiber	*1.4g*	★

HAUNCH OF VENISON WITH WILD MUSHROOMS

Venison, like all game, is very lean, and used to be heavily larded and barded before cooking. Indeed, it is hard to make venison tasty unless it is either larded or marinated.
The latter course is, naturally, the most healthy and best suited to contemporary tastes.
The best varieties of wild mushrooms to use with venison are morels, chanterelles or cèpes (often sold under the Italian name of porcini). If you are using fresh wild mushrooms there is no need to soak them.
When venison is not available any kind of game can be cooked in this way.

· METHOD ·

Combine the marinade ingredients in a bowl or pot (not one made of aluminum) and bring to the boil. Allow to cool to room temperature. Add the venison to the marinade and leave it overnight.

When ready to cook, drain the meat, reserving the marinade. Strain the marinade into a casserole; add the venison and the wild mushrooms, together with their soaking water if using dried. Season with salt and add the juniper berries and peppercorns.

Cover the casserole and cook in a preheated 325°F oven for 3 hours, or until the meat is tender.

Remove the venison from the casserole and brown it quickly under a hot broiler. Stir the cornstarch into the yogurt with a tablespoon of the cooking juices. Stir the yogurt mixture into the cooking juices and pour the sauce over the venison just before serving.

· SERVING SUGGESTION ·

Serve with steamed potato balls or mashed potatoes. Bean sprouts and alfalfa sprouts make an unusual but harmonious vegetable accompaniment. You can also add watercress or garden cress and lettuce for garnish. Cranberry sauce or Cumberland sauce should be passed around as a relish.

ROAST PHEASANT WITH BLACKBERRY AND APPLE

Pheasant is another very lean game bird that must be roasted with some fat to keep it moist.
Use bacon, which is much leaner than barding fat, and discard it when the meat is cooked. Because of the fattiness of the resulting cooking juices, they are not incorporated into the sauce. But the juices can be chilled and any fat skimmed off and discarded before the liquid is used for another dish. If blackberries are not available, blueberries, loganberries or similar berries can be substituted.

· METHOD ·

Put the neck and giblets into a saucepan with the broth and Calvados or applejack. Cover the pan and simmer for 30 minutes. Add the apple slices. Cook for another 15 minutes. Preheat the oven to 450°F.

Arrange 3 bacon slices over each bird and truss them in place. Place the birds, breast downward, on a rack over a roasting pan and roast them for 20 minutes. Turn them over and roast for another 15 minutes.

While the birds are cooking, strain and reserve the liquid from the saucepan. Discard the necks and gizzards – or use them in another dish, such as Cajun Rice (see page 171) – and reserve the livers and the apple. Finely chop the livers and apple in a food processor to a smooth paste.

Toast the bread slices on both sides. Remove the crusts and cut carefully through the center of each slice to make two thin Melba toasts; briefly toast again. Arrange the liver-and-apple mixture in a small dish with the Melba toast cut into triangles and arranged around it.

Remove the birds from the oven and transfer them to a large serving dish. Add ¾ cup of the berries to the strained liquid and return it to the pan. Cook just long enough to heat through. Pour some of the liquid over the birds and pass the rest separately. Decorate with the remaining uncooked berries.

· SERVING SUGGESTION ·

Red cabbage, cooked with red wine vinegar, honey and golden raisins, makes an excellent accompaniment to this delicious game dish. Roast or boiled potatoes would also go well with it. Remember that very slow boiling will make potatoes taste much nicer; boiled potatoes should never be reheated or they will taste stale.

INGREDIENTS

2 young hen pheasants, drawn, wingtips discarded, neck removed
1 cup chicken broth (page 24)
¼ cup Calvados or applejack
2 large tart apples, cored and sliced
6 Canadian-style bacon strips
1 cup blackberries or blueberries
8 slices whole wheat bread

8 servings
Picture: page 159

Calories	259	★ ★
Fat	7g	★ ★ ★
Saturated Fat	2.3g	★ ★
Cholesterol	14mg	★ ★ ★
Sodium	313mg	★ ★
Fiber	6.2g	★ ★

SADDLE OF RABBIT WITH PRUNES AND APRICOTS

for recipe see page 165

ROAST PIGEON WITH BUCKWHEAT GROATS

for recipe see page 164

INGREDIENTS

1 tablespoon honey

1 tablespoon soy sauce

2 tablespoons chopped celery leaves

*2 pigeons, drawn, livers reserved
and chopped*

2oz pine nuts (pignolas)

8oz buckwheat (kasha)

1 egg, beaten

*½ cup dried mushrooms, soaked
in water*

2 cups chicken broth (page 24)

2 servings

Picture: page 163

Calories	*687*	★
Fat	*31g*	★
Saturated Fat	*1.4g*	★ ★ ★
Cholesterol	*187mg*	★ ★
Sodium	*393mg*	★ ★
Fiber	*4.2g*	★ ★

INGREDIENTS

*2 rabbit saddles and backs
(about 3lb total)*

*1 cup chicken or mixed
meat broth (page 25)*

½ cup whole wheat breadcrumbs

1 garlic clove, chopped

1 medium onion, coarsely chopped

½ teaspoon black pepper

1 tablespoon chopped parsley

1 teaspoon dried marjoram

1 tablespoon Dijon-style mustard

½ cup soy bacon-flavored bits

1 tablespoon brandy

2 egg whites

6 servings

ROAST PIGEON WITH BUCKWHEAT GROATS

**The strong flavor of pigeon or squab goes well with
the nutty taste of buckwheat. Quail can be cooked in the same way.
Both meats are rich in flavor but low in fat.**

· METHOD ·

Mix the honey, soy sauce and celery leaves. Brush the pigeons inside and out with the mixture. Sprinkle half the pine nuts inside the birds and arrange them on a rack. Grill them, breast side down, for 10 minutes, then turn over for another 10 minutes. Alternatively, the pigeons can be roasted for 10 minutes on either side in a preheated 350°F oven.

While the birds are cooking put the buckwheat into a dry skillet over low heat. Add the beaten egg and stir well so that the grains absorb the egg. Add the mushrooms in their soaking liquid and the broth and cover the pan. Simmer for 10 minutes. Stir to fluff the grains, then cover and simmer again for 5 minutes.

Arrange the buckwheat on a warmed serving dish and place the pigeons on top. Pour any remaining basting liquid over the top and sprinkle with the remaining pine nuts.

· SERVING SUGGESTION ·

A suitable accompaniment to this strongly flavored dish is a green salad, especially one containing bitter greens, such as curly endive or chicory. Any leftover giblets can be used with Cajun Rice (see page 171).

RABBIT PATE

**Normally, pâté is very heavily larded with fat.
This lean version tries to achieve moistness by other means.
A microwave, combination or steam oven achieves the best results.
The conventional cooking method is given here, but if you are using
an alternative, cook the assembled pâté, covered in plastic wrap, on
high for 8 minutes, turning off the oven briefly once
halfway through.**

· METHOD ·

Heat the oven to 350°F. Put the rabbit into a large deep pan and add the broth. Cover the pan tightly and cook the rabbit for 45 minutes. Remove it from the cooking liquid (which can be used in another dish) and cut the meat off the bone. Cut half the meat into bite-sized pieces and reserve.

Put the rest into a food processor and grind it with the breadcrumbs, garlic, onion, pepper, parsley, marjoram and mustard. Transfer the mixture to a bowl and stir the bacon-flavored bits into it. Whip the egg whites into stiff peaks and fold them little by little into the mixture. Line a nonstick loaf pan with nonstick baking paper, cutting it to fit and making sure it is not creased, as the creases will show on the pâté. Divide the ground rabbit mixture in half. Spread one half evenly over the bottom of the pan, then arrange the bite-sized pieces of rabbit on top. Cover with the rest of the ground mixture, pressing it down well. Cover the loaf pan tightly with foil, pressing it down well at the sides to ensure there are no gaps. Place the pan in a pan of warm water to come two-thirds of the way up the sides. Bake in a preheated 325°F oven for 1 hour. It will begin to come away from the sides of the pan when done.

If a browned crust is desired, remove the foil for the last 10 minutes of cooking. Unmold, peel away the paper and leave to cool. Refrigerate when cold. Decorate the top with raw vegetables before serving.

· SERVING SUGGESTION ·

The pâté can be eaten either as a starter or as a main course with salad and whole wheat bread. It also makes a very elegant buffet dish, or a picnic treat, stuffed into whole wheat pita.

Calories	300	★ ★
Fat	7g	★ ★ ★
Saturated Fat	2.1g	★ ★
Cholesterol	110mg	★ ★
Sodium	382mg	★ ★
Fiber	2.4g	★

SADDLE OF RABBIT WITH PRUNES AND APRICOTS

Rabbit is too often overlooked by those who care about healthy eating. It is inexpensive and very low in fat, with a firm flesh rather like chicken. The alcohol used in this recipe evaporates during cooking, leaving a delicious rum flavor behind.

INGREDIENTS

1 saddle and hindquarters of
   rabbit, cut into serving pieces
3 cups pitted prunes
3 cups dried apricots
5 tablespoons dark rum
Juice and grated rind of 1 lime
1 sprig thyme
1 bay leaf
2 tablespoons oil
2 tablespoons wholewheat flour
2 cups chicken broth (page 24)

4 servings
Picture: page 162

· METHOD ·

Marinate the rabbit pieces in the prunes, apricots, rum, lime juice and rind, thyme and bay leaf overnight or for up to 24 hours, turning occasionally.

Drain the rabbit pieces thoroughly, but reserve the marinade. Heat the oil in a nonstick skillet and brown the rabbit pieces, sprinkling them with the flour as they cook. When the flesh is firm, remove from the skillet and transfer them to a casserole. Pour the marinade over and add the chicken broth. Cover tightly. Simmer over low heat for 45 minutes, or until the rabbit is tender. (Alternatively, the dish can be microwaved on medium for 20 minutes.) Discard the thyme and bay leaf before serving.

· SERVING SUGGESTION ·

This hearty meat meal deserves a hearty vegetable accompaniment, such as puréed potatoes and turnips. Red cabbage, shredded and steamed with a little cider vinegar and a tablespoon of honey, also complements the dish.

Calories	619	★
Fat	12g	★ ★
Saturated Fat	2.9g	★ ★
Cholesterol	80mg	★ ★
Sodium	272mg	★ ★
Fiber	41.0g	★ ★ ★

CAJUN RICE
for recipe see page 171

SWEET AND SOUR TONGUE
for recipe see page 172

VARIETY MEATS

Variety meats can deteriorate quickly. Always look for pieces that are fresh and moist looking. Liver and kidney should be bright to dark red. Unwrap the variety meat, rewrap it loosely in paper and refrigerate it immediately in the coldest part of the refrigerator. Liver and heart can be frozen but must be allowed to thaw quickly. Other types of variety meats cannot be frozen.

Always cook fresh variety meats within 24 hours of purchase. Variety meats such as brains and sweetbreads must be soaked, peeled and poached lightly in boiling water before any other preparation. Allow 8oz per serving portion. Liver and kidneys can be left faintly pink on the inside, but other kinds of variety meats should be cooked very thoroughly, especially tripe.

BROILED LIVER WITH FRESH HERBS

Almost all recipes for roasting liver call for the liver to be larded
with pieces of pork fat or bacon.
This is not necessary if the liver is first grilled quickly, sealing in the
juices while keeping the inside moist. There is no need even to flour
the liver first, but it needs just a light brush of oil or margarine or
the outside will become tough and leathery as the inside cooks.
The meat should be still pink in the center when properly cooked.
Liver is the most nutritious meat for its weight, being especially
rich in iron and vitamin B. As the meat juices are not fatty, they can
be collected and used for gravy.
Ox or beef liver is rather coarse-grained and particularly
strong-flavored, with lots of connective tissue.
Lambs' liver is twice as expensive but much more delicate. Most
expensive of all, but by far the tastiest, is calves' liver. Never season
liver until it has been cooked.

· METHOD ·

Cut 4 small slits in the liver and place the herbs in the slits.

Place the liver on a rack over a metal pan placed in the broiler tray to
catch the drips. Pour $1/4$ cup of water into the pan. Heat the broiler to hot.
Broil the liver quickly for 1 minute on each side; brush the surfaces with
oil, then return to the broiler and cook for 3 minutes more on each side, or
until well browned. Slice thinly to serve. Mix the water and drippings in
the pan under the broiler with the wine, and pour it over the liver. Arrange
the watercress in sprigs around the liver.

· SERVING SUGGESTION ·

The distinctive flavor of liver is best offset by a bland, starchy vegetable.
Instead of the usual potato, try polenta - cornmeal boiled while being
stirred continuously - or simply heated canned corn. If you are fond of
traditional liver and onions, stew some sliced onions in broth to cover for
15 minutes or until tender, and serve with the meat.

INGREDIENTS

1lb liver, in 1 or 2 pieces

1 tablespoon fresh sage or
    rosemary

2 bay leaves

1 tablespoon oil

$1/4$ cup dry red wine

1 bunch watercress

4 servings

Calories	247	★ ★ ★
Fat	15g	★ ★
Saturated Fat	4.1g	★ ★
Cholesterol	484mg	★
Sodium	95mg	★ ★ ★
Fiber	0.4g	★

INGREDIENTS

1 tablespoon oil

1 onion, grated

*8oz chicken livers, trimmed of
any green stains and
connective tissue*

2 tablespoons brandy

*½ teaspoon salt or salt substitute
(page 26)*

½ teaspoon black pepper

1 teaspoon chopped mixed herbs

5oz skimmed milk cheese

3 tablespoons low-fat milk

4 small servings

Calories	*174*	★ ★ ★
Fat	*9g*	★ ★
Saturated Fat	*1.6g*	★ ★ ★
Cholesterol	*240mg*	★
Sodium	*322mg*	★ ★
Fiber	*0.5g*	★

INGREDIENTS

1lb spinach

1lb chicken livers

1 cup milk

*1 tablespoon margarine or oil for
greasing the dish*

3 tablespoons grated Parmesan cheese

4 servings

LIGHT CHICKEN LIVER PATE

**All liver is highly nutritious, being very rich in iron and vitamin A,
but due to its strong flavor it is not always popular. When the flavor
is combined with other ingredients to make it less overpowering,
however, it is quite delicious. Unfortunately, this is done mostly by
combining the liver with cream! Salvation is at hand, though: the
same effect can be achieved by using low-fat cheese. The best for
the purpose is a skimmed milk cheese; if you cannot find it, use a
low-fat soft cheese or tofu.**

· METHOD ·

Heat the oil in a nonstick skillet. Add the grated onion and the livers and
cook on high heat, stirring occasionally, until the livers are no longer
brown on the outside. Pour the brandy into a warmed spoon and tip it into
the pan. Set it alight with a match and shake the livers into it.

Transfer the mixture to a food processor and grind until smooth. Add
the salt or salt substitute, peppers, herbs and the skimmed milk cheese,
and continue grinding. While the machine is running, add the milk. Stop
the machine and check that the mixture is smooth. If it seems too dry, you
can add another tablespoon or two of milk while processing. The amount
of milk required will depend on the consistency of the cheese.

Shape the mixture into a neat mound, and chill before serving.

· SERVING SUGGESTION ·

Serve a little mound of the pâté as a first course with salad or with whole
wheat biscuits, bread or crackers as a dip or appetizer. The pâté makes an
excellent picnic spread or buffet dish, especially if pressed into a wetted
fancy mold and turned out. Brush it with egg white to give it an attractive
shiny glaze.

CHICKEN LIVER AND SPINACH GRATIN

**This dish is very nutritious, being particularly rich in iron.
It is cooked conventionally in this recipe, but it is also ideal for
microwaving up to the cheese stage, in which case the dish will not
need to be greased. Microwave it in a glass dish on high for 20
minutes, without adding the cheese. Add the cheese, then finish
under a hot broiler.**

· METHOD ·

Rinse and trim the spinach leaves. Drop them into a dry saucepan and
cook over medium heat for 2 minutes, or until the spinach is wilted.
Remove and leave to cool while you finely mince the chicken livers with a
knife or grind them in a food processor.

Chop the cooked spinach finely and combine it with the chicken livers. Moisten the mixture with the milk. Pat it into a well-greased baking dish and sprinkle with the cheese. Bake in a preheated 350°F oven for 45 minutes, or until the cheese has melted and is golden.

· SERVING SUGGESTION ·

Rice, either white or brown, is the obvious accompaniment to this strongly flavored dish. A tomato salad with chives would make a refreshing vegetable.

Calories	280	★ ★
Fat	15g	★ ★
Saturated Fat	5.3g	★
Cholesterol	440mg	★
Sodium	320mg	★ ★
Fiber	0.6g	★

CAJUN RICE

**This traditional recipe comes from the Cajun country of Louisiana. The first Cajun settlers came from Provence, and their cooking reflects its Provençal roots. The original recipe uses plenty of pork fat, but as usual, this was probably just to grease the pan so the food didn't stick.
The flavor is actually improved without it. This is also an excellent way to use up any meat leftovers, in addition to giblets.**

INGREDIENTS

1lb chicken gizzards
3 cups chicken broth (page 24)
8oz ground pork loin
½ cup chopped onion
3 celery stalks, chopped
1 green bell pepper, seeded and chopped
3 green onions, chopped
2 garlic cloves, chopped
1 cup long-grain brown rice
2 teaspoons Creole Spice (page 27)
8oz chicken livers, coarsely chopped

8 servings

Picture: page 166

· METHOD ·

In a saucepan simmer the gizzards in the broth for 30 minutes. Remove them with a slotted spoon and chop them finely. Reserve the broth. Put the ground pork into a nonstick skillet with the gizzards and cook over high heat until browned. Reduce the heat and add the vegetables. Cook for 5 minutes or until the onion is transparent, stirring constantly.

Add the rice to the broth and cover the saucepan. Cook for 15 minutes. Drain the contents of the skillet, reserving any fat. Add the gizzards and vegetables to the saucepan with the Creole Spice and cook for another 15 minutes. Meanwhile, put the livers into the skillet and sauté them briefly until browned on the outside. Add them to the rice and cook, uncovered, stirring frequently for 10 minutes or until most of the liquid is absorbed.

· SERVING SUGGESTION ·

This dish is a complete meal in itself but would be enhanced by a salad, for example, a tomato salad sprinkled with chopped fresh herbs including green onions or chives. Large tomatoes can be sliced crosswise, cherry tomatoes merely cut in half. Sprinkle with lemon juice or vinegar.

Calories	253	★ ★
Fat	8g	★ ★
Saturated Fat	2.7g	★ ★
Cholesterol	357mg	★
Sodium	376mg	★ ★
Fiber	1.9g	★

INGREDIENTS

6 fresh lambs' or calves' tongues (about 6oz each)

2 garlic cloves

2 tablespoons pickling spice, tied in cheesecloth

1 cup cider vinegar

3 tablespoons honey

½ cup golden raisins

6 ginger snaps, crushed or crumbled

6 servings

Picture: page 167

Calories	432	★ ★
Fat	27g	★
Saturated Fat	8.6g	★
Cholesterol	315mg	★
Sodium	779mg	★
Fiber	1.0g	★

SWEET AND SOUR TONGUE

**The crumbling of gingerbread into a sauce to thicken it goes right back to the Romans, who used pastry in this way.
The gingerbread used in the cooking of eastern Germany and Poland is almost like a rusk made with pepper and honey, even harder than our gingersnaps.
The pickled version of tongue is rather salty. It is a better idea to make use of fresh tongue when available. It is surprisingly little trouble to cook, and there is hardly any waste or surplus fat. If you cook a whole ox tongue this way, simply increase the simmering time to 2 hours.**

· METHOD ·

Trim the fatty parts from the base of the tongues and remove the cartilage. Put the tongues into a deep pot and cover with cold water. Bring to the boil and cook on medium heat, uncovered, for 10 minutes, frequently skimming the surface.

Reduce the heat to a bare simmer and stir in the garlic cloves and pickling spice. Add more water if the tongues are uncovered, and put a lid on the pot. Simmer for 2 hours.

Remove the tongues from the liquid. Strain the liquid through a sieve and pour it back into the pan. Skin the tongues. Return the skinned tongues to the pan and add the vinegar, honey, raisins and ginger snaps. Cover the pot and simmer for 30 minutes.

Remove and drain the tongues, and transfer them to a carving board. If the sauce has not thickened, increase the heat and boil it, uncovered, for 10 to 15 minutes, or until it is reduced. Meanwhile, slice the tongues neatly, pressing them back into shape. Put them on a serving dish in a warm place. Pour some of the sauce over them and serve the rest in a sauce boat.

· SERVING SUGGESTION ·

Steamed or boiled cauliflower or broccoli makes an attractive and tasty combination with tongue. If you break the vegetables into flowerets, they can be steamed quite quickly. Turnip greens and Swiss chard are strong-flavored green vegetables that would also go well with this dish.

TONGUE SALAD WITH PICKLED VEGETABLES AND MUSTARD SAUCE

Tongue is red meat and higher in fat than poultry. But small amounts go a long way, and as it is usually eaten cold, here is a good and original idea for a tasty salad. Pickled red cabbage is quite easy to find nowadays, but if it is unavailable, combine two cooked, chopped beets with well-rinsed and drained sauerkraut for a similar effect and flavor.

· METHOD ·

Mix the tongue, potato, apple, pickled cabbage and sour pickles. Have ready 4 salad plates. Lay a cabbage leaf on each and fill the leaf with a quarter of the mixture. Sprinkle the chopped fresh coriander over the salad.

To make the mustard sauce, combine all the ingredients in a food processor or blender. Serve separately or on the side.

· SERVING SUGGESTION ·

This salad is a balanced meal in itself but some people would prefer a little more starch with it. This can be provided in the form of more boiled potatoes – preferably new ones – served with low-fat yogurt.

INGREDIENTS

8oz cooked tongue, diced

1 cup cold boiled potato, diced

1 large tart apple (about 5oz), cored and diced

1½ cups pickled red cabbage, coarsely chopped

1 sour pickle, diced

4 small inner red cabbage leaves, trimmed into rounds

4 tablespoons chopped fresh coriander

MUSTARD SAUCE

2 tablespoons Dijon-style mustard

1 garlic clove, chopped

1 teaspoon brown sugar

1 tablespoon cider vinegar

¼ cup tofu

4 servings

Calories	274	★ ★
Fat	17g	★
Saturated Fat	7.0g	★
Cholesterol	63mg	★ ★ ★
Sodium	652mg	★
Fiber	4.2g	★ ★

INDEX

Page numbers in *italic* refer to the illustrations

ACKNOWLEDGMENT

Quarto Publishing plc would like to thank
Abbie Sinclair of Elizabeth David Ltd.
and the other retailers who supplied kitchen equipment
used in the recipe photographs in this book.